网络空间安全系列教材

计算机网络安全

主　编　陈显毅
副主编　周晓谊

电子工业出版社
Publishing House of Electronics Industry
北京·BEIJING

内 容 简 介

本书从计算机网络安全原理、交换机安全实践、路由器安全和防火墙安全 4 个层面介绍计算机网络安全及实践的知识，内容包括网络犯罪与攻击方法、密码学、防火墙技术、安全远程登录网络设备、MAC 地址表及端口安全、多出口主机网关安全、IPv6 邻居发现协议、IPv6 无状态地址自动配置、DHCPv6——地址自动分配、基于路由器的 CBAC 防火墙、基于路由器的 IPsec VPN、基于路由器的 Easy VPN、ASA 防火墙的基本配置、ASA 防火墙的路由配置、ASA 防火墙的 NAT 和 ACL 配置、策略路由和基于 ASA 防火墙的 IPsec VPN 配置等。

本书可作为高等院校网络安全、信息安全、密码科学、计算机科学与技术等相关专业计算机网络安全类课程的配套教材，也可作为从事网络安全工作的工程技术人员的参考书。

未经许可，不得以任何方式复制或抄袭本书之部分或全部内容。
版权所有，侵权必究。

图书在版编目（CIP）数据

计算机网络安全 / 陈显毅主编. -- 北京 : 电子工业出版社, 2025. 6. -- ISBN 978-7-121-50555-3

Ⅰ．TP393.08

中国国家版本馆 CIP 数据核字第 20254Z0S03 号

责任编辑：牛晓丽
印　　刷：北京雁林吉兆印刷有限公司
装　　订：北京雁林吉兆印刷有限公司
出版发行：电子工业出版社
　　　　　北京市海淀区万寿路 173 信箱　　　　邮编：100036
开　　本：787×1092　1/16　　印张：15.25　　字数：400 千字
版　　次：2025 年 6 月第 1 版
印　　次：2025 年 6 月第 1 次印刷
定　　价：59.80 元

凡所购买电子工业出版社图书有缺损问题，请向购买书店调换。若书店售缺，请与本社发行部联系，联系及邮购电话：(010) 88254888，88258888。
质量投诉请发邮件至 zlts@phei.com.cn，盗版侵权举报请发邮件至 dbqq@phei.com.cn。
本书咨询联系方式：QQ 9616328。

前言

本书首先介绍了计算机网络安全的基本原理，在此基础上，系统讨论了交换机和路由器的安全实践。全书共 17 章，第 1 章概述网络犯罪与攻击方法；第 2 章介绍古典加密算法、对称加密算法和非对称加密算法；第 3 章介绍防火墙的概念、防火墙的分类、防火墙的结构、防火墙的部署和包过滤防火墙的配置；第 4 章介绍安全远程登录网络设备的组网需求和仿真实验；第 5 章介绍交换机 MAC 地址表、端口安全和仿真实验；第 6 章介绍多出口主机网关安全，通过路由冗余协议提高网络可靠性，同时实现负载均衡；第 7 章介绍通过 IPv6 邻居发现协议安全地实现 MAC 地址解析、重复地址检测和邻居状态跟踪等功能；第 8 章介绍 IPv6 主机通过无状态地址自动配置方式从路由器安全地获得全球 IPv6 地址、默认网关、跳数限制等网络参数；第 9 章介绍 DHCPv6 客户端通过有状态地址自动分配方式从 DHCPv6 服务器安全地获取 IPv6 地址、DNS 服务器地址等网络参数；第 10 章介绍路由器通过 CBAC（Context-Based Access Control，基于上下文的访问控制）实现网络资源的安全访问，使得安全等级高的网络区域可以访问安全等级低的网络区域内的任何资源，同时有效抵御拒绝服务攻击；第 11 章介绍在两个安全网关之间建立站点到站点的 IPsec VPN，实现公司分部网段内的用户可以安全访问公司总部网络的内部资源；第 12 章介绍借助 Easy VPN 网关，出差到外地的公司员工通过互联网拨号接入 Easy VPN 服务器，从而安全访问公司内部网络资源；第 13 章介绍防火墙的安全区域、安全等级、安全远程登录等基本配置；第 14 章介绍防火墙的静态路由、默认路由、动态路由、路由导入等路由操作；第 15 章介绍防火墙的网络地址转换、端口地址转换、Identity NAT 和不同安全等级区域间的访问；第 16 章介绍通过策略路由实现基于源 IP 地址的路由，以满足负载均衡的需求，同时提高内部网络访问外部网络的可靠性和安全性；第 17 章介绍在两个防火墙之间配置站点到站点的 IPsec VPN，使分公司的用户可以通过浏览器安全访问总公司的内部 Web 服务器。

全书由陈显毅博士主持编写和统稿。第 1 章至第 5 章由周晓谊博士编写，第 6 章至第 15 章由陈显毅编写，第 16 章和第 17 章由吴泽群编写。

本书的编写工作得到了海南大学研究生自编教材、海南大学教育教学改革研究项目（hdjy2053 和 hdjy2125）、海南省高等学校教育教学改革研究项目（Hnjg2021-7）的立项支持，电子工业出版社也提供了积极的协助，在此一并致以诚挚的谢意！

由于编者水平有限，书中难免会有疏漏和错误，恳请读者批评和指正。如有任何问题，欢迎直接通过邮件与编者联系：chenxianyi@hainanu.edu.cn。

编者

2025 年 5 月

目录

第1篇 计算机网络安全原理篇

第1章 网络犯罪与攻击方法 ... 2
- 1.1 网络犯罪概述 ... 2
- 1.2 实施网络犯罪的方法 ... 2
  - 1.2.1 缓冲区溢出攻击 ... 3
  - 1.2.2 拒绝服务攻击 ... 8
  - 1.2.3 网络犯罪的特点 ... 10
  - 1.2.4 网络犯罪者 ... 11
  - 1.2.5 黑客拓扑结构 ... 11
  - 1.2.6 黑客攻击发展趋势 ... 14
- 本章小结 ... 14

第2章 密码学 ... 16
- 2.1 基础知识 ... 16
  - 2.1.1 可能的攻击 ... 17
  - 2.1.2 单向函数与单向 Hash 函数 ... 17
  - 2.1.3 对称密码和非对称密码 ... 18
  - 2.1.4 密钥长度 ... 19
  - 2.1.5 密码学应用 ... 20
- 2.2 古典密码学 ... 21
  - 2.2.1 凯撒密码 ... 21
  - 2.2.2 希尔密码 ... 22
  - 2.2.3 栅栏密码 ... 23
  - 2.2.4 古典密码安全性分析 ... 23
- 2.3 分组密码 DES ... 24
  - 2.3.1 DES 轮迭代 ... 25
  - 2.3.2 DES 密钥生成及安全性 ... 26
- 2.4 分组密码 AES ... 26
  - 2.4.1 AES 轮迭代 ... 27
  - 2.4.2 轮密钥生成器 ... 29
  - 2.4.3 AES 算法实现 ... 30

2.5　非对称密码体制 31

第 3 章　防火墙技术 32
　　3.1　防火墙概述 32
　　　　3.1.1　防火墙的基本概念 32
　　　　3.1.2　防火墙的功能 33
　　3.2　防火墙的分类 33
　　　　3.2.1　分组过滤型防火墙 33
　　　　3.2.2　应用代理型防火墙 33
　　　　3.2.3　复合型防火墙 34
　　3.3　防火墙的结构 34
　　　　3.3.1　堡垒主机 34
　　　　3.3.2　主机驻留防火墙 35
　　　　3.3.3　网络设备防火墙 35
　　　　3.3.4　个人防火墙 35
　　3.4　防火墙的部署 36
　　　　3.4.1　DMZ 的部署 36
　　　　3.4.2　VPN 的部署 38
　　　　3.4.3　分布式防火墙的部署 39
　　3.5　包过滤防火墙的配置 41
　　　　3.5.1　ACL 概念 41
　　　　3.5.2　ACL 配置命令 42
　　　　3.5.3　ACL 的应用 43

第 2 篇　交换机安全实践篇

第 4 章　安全远程登录网络设备 46
　　4.1　组网需求 46
　　4.2　仿真实验 46
　　　　4.2.1　仿真环境 46
　　　　4.2.2　实验设计 47
　　　　4.2.3　操作步骤 49

第 5 章　MAC 地址表及端口安全 56
　　5.1　组网需求 56
　　5.2　仿真实验 57
　　　　5.2.1　仿真环境 57
　　　　5.2.2　实验设计 57
　　　　5.2.3　操作步骤 58

第 6 章　多出口主机网关安全 62
　　6.1　组网需求 62
　　6.2　仿真实验 63

		6.2.1 仿真环境	63
		6.2.2 实验设计	63
		6.2.3 操作步骤	64

第 7 章　IPv6 邻居发现协议 ... 75

7.1　组网需求 ... 75
7.2　仿真实验 ... 76
  7.2.1　仿真环境 ... 76
  7.2.2　实验设计 ... 76
  7.2.3　操作步骤 ... 78

第 3 篇　路由器安全篇

第 8 章　IPv6 无状态地址自动配置 ... 102

8.1　组网需求 ... 102
8.2　仿真实验 ... 103
  8.2.1　仿真环境 ... 103
  8.2.2　实验设计 ... 103
  8.2.3　操作步骤 ... 104

第 9 章　DHCPv6——地址自动分配 ... 117

9.1　组网需求 ... 117
9.2　仿真实验 ... 118
  9.2.1　仿真环境 ... 118
  9.2.2　实验设计 ... 118
  9.2.3　操作步骤 ... 119

第 10 章　基于路由器的 CBAC 防火墙 ... 131

10.1　组网需求 ... 131
10.2　仿真实验 ... 132
  10.2.1　仿真环境 ... 132
  10.2.2　实验设计 ... 132
  10.2.3　操作步骤 ... 133

第 11 章　基于路由器的 IPsec VPN ... 143

11.1　组网需求 ... 143
11.2　仿真实验 ... 143
  11.2.1　仿真环境 ... 143
  11.2.2　实验设计 ... 144
  11.2.3　操作步骤 ... 145

第 12 章　基于路由器的 Easy VPN 157
12.1　组网需求 157
12.2　仿真实验 157
12.2.1　仿真环境 157
12.2.2　实验设计 158
12.2.3　操作步骤 159

第 4 篇　防火墙安全篇

第 13 章　ASA 防火墙的基本配置 170
13.1　组网需求 170
13.2　仿真实验 170
13.2.1　仿真环境 170
13.2.2　实验设计 171
13.2.3　操作步骤 172

第 14 章　ASA 防火墙的路由配置 185
14.1　组网需求 185
14.2　仿真实验 185
14.2.1　仿真环境 185
14.2.2　实验设计 186
14.2.3　操作步骤 187

第 15 章　ASA 防火墙的 NAT 和 ACL 配置 194
15.1　组网需求 194
15.2　仿真实验 195
15.2.1　仿真环境 195
15.2.2　实验设计 195
15.2.3　操作步骤 196

第 16 章　策略路由 208
16.1　组网需求 208
16.2　仿真实验 209
16.2.1　仿真环境 209
16.2.2　实验设计 209
16.2.3　操作步骤 210

第 17 章　基于 ASA 防火墙的 IPsec VPN 配置 225
17.1　组网需求 225
17.2　仿真实验 226
17.2.1　仿真环境 226
17.2.2　实验设计 226
17.2.3　操作步骤 227

第 1 篇　计算机网络安全原理篇

第 1 章　网络犯罪与攻击方法

第 2 章　密码学

第 3 章　防火墙技术

第 1 章 网络犯罪与攻击方法

当今世界，以信息技术为首的高科技形成了一股前所未有的科技浪潮，将人类社会带入了高科技网络时代。爱因斯坦曾用一段话来提醒人们重视科学技术的双刃性："以前几代人给了我们高度发展的科学与技术，这是一份最宝贵的礼物，它使我们有可能生活得比以前无论哪一代人都要自由和美好。但是，这份礼物也带来了从未有过的巨大危险，它威胁着我们的生存。"事实上，任何事物都具有两面性，网络在为人类带来便捷的同时，也为新犯罪形态的演化开辟了道路。

1.1 网络犯罪概述

随着网络的日益普及，网络犯罪也日渐增多与复杂化，犯罪分子在互联网上开辟了新的犯罪平台，它所影响层面的广度与深度、所造成的危险与侵害都是人类社会史无前例的。网络犯罪已经对社会构成了现实的威胁，严重威胁着互联网的发展，直接影响着国家政治、经济、文化等方面的正常秩序，成为信息时代的最大隐患。

美国国家基础设施保护中心（NIPC）的一位主管曾说："网络犯罪总会对电子商务和公众造成极大的威胁。"利用互联网进行犯罪的威胁是真实的、不断增长的，并且很可能是 21 世纪苦难的根源。网络犯罪同其他犯罪一样，只是必须涉及计算机系统，计算机系统既可以作为犯罪的对象，也可以作为实施犯罪的设备或者犯罪证据库。网络犯罪是指行为人运用计算机技术，借助网络对系统或信息进行攻击，破坏或利用网络进行犯罪的总称，既包括行为人运用编程、加密、解码技术或工具在网络上实施的犯罪，也包括行为人利用软件指令、网络系统、产品加密等技术及法律的漏洞在网络内外交互实施的犯罪，还包括行为人借助其居于网络服务提供者特定地位或其他方法在网络系统中实施的犯罪。简而言之，网络犯罪是针对或利用网络进行的犯罪，其本质特征是危害网络及其信息的安全与秩序。

1.2 实施网络犯罪的方法

网络犯罪的目的是破坏网络信息的保密性、完整性、可审查性，破坏网络服务的可用性，破坏网络运行的可控性。任何网络犯罪都必须借助计算机资源才能顺利实施，常见的网络犯罪方法包括信息探测攻击、密码破解攻击、远程控制攻击、缓冲区溢出攻击、拒绝服务攻击、嗅探和欺骗攻击等。下面对两类典型的网络犯罪方法——缓冲区溢出攻击和拒绝服务攻击展开讨论。

1.2.1 缓冲区溢出攻击

最早的缓冲区溢出攻击可以追溯到 1988 年的 Morris 蠕虫，该蠕虫利用 fingerd 进程存在的缓冲区漏洞实施攻击。在此之后，各种缓冲区溢出漏洞相继被发现与利用，在全球范围内造成了极大的损失。比较典型的有：2001 年，红色代码（code red）蠕虫利用微软 IIS 5.0 的一个缓冲区溢出漏洞获得了超级用户权限，以实施进一步的网络攻击。2003 年，Slammer 蠕虫利用了微软 SQL Server 2000 的一个缓冲区溢出漏洞，该蠕虫主要攻击基于 SQL Server 2000 的服务器，使之不能正常工作，造成了巨大的负面影响。2004 年，"震荡波"蠕虫利用 Windows 2000/XP 的本地安全授权子系统服务（Local Security Authority Subsystem Service，LSASS）缓冲区溢出漏洞实施攻击。该蠕虫一旦攻击成功，就会感染主机并快速在网络中传播；该蠕虫若攻击失败，也会造成目标主机的缓冲区溢出，导致目标主机因出现内存非法操作等异常而停机。

1.2.1.1 缓冲区溢出攻击原理

为了更好地说明缓冲区溢出攻击的原理，下面先介绍计算机内存通常采用的地址空间布局。32 位 Linux 操作系统的内存地址空间布局如图 1-1 所示，其中内核区位于内存的高端区域，c0000000～ffffffff 一共 1GB 的空间分配给操作系统内核。栈区位于内核区的下方，其大小是动态分配的，具体来说就是从高地址端向低地址端分配。栈区主要有以下两个作用：一是动态存储函数之间的调用与被调用关系，以保证被调用函数在返回到主函数时正确执行后续命令；二是为函数的局部变量提供存储空间。为了节省系统资源，动态链接库是动态加载的，操作系统会把必要的动态链接库加载到内存中。堆区在数据区的上方，主要用于进程的临时变量动态申请一定大小的内存空间，其地址分配方式与栈区相反，即从内存低地址端向高地址端依次分配。数据区用于存储进程的全局变量和常量。指令区用于存放程序指令，待执行的二进制机器代码在该区域中载入，处理机会到这个区域中取指令并译码执行。

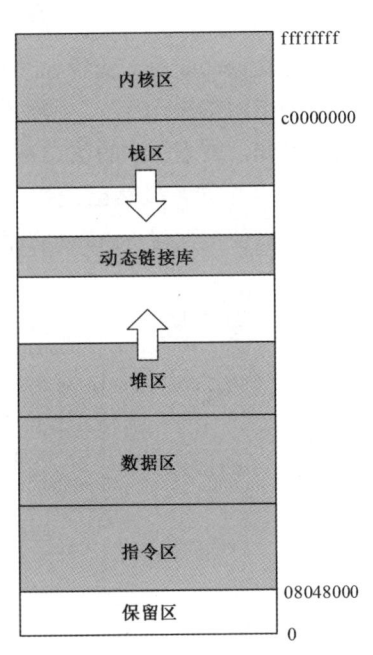

图 1-1　内存地址空间布局

缓冲区可设置在进程的栈区、堆区和数据区，一旦发生缓冲区溢出，就会引发以下后果：程序使用的数据遭到破坏、程序控制权非法转移、内存非法访问和程序异常终止。本节讨论缓冲区溢出导致数据遭到破坏，程序控制权非法转移等将在后面展开分析。

下面以一个简单的存在漏洞的 C 语言程序（见图 1-2）为例说明缓冲区溢出的基本过程，以及缓冲区溢出是如何导致数据遭到破坏的。图 1-2 的第 6 行代码中存在缓冲区溢出漏洞，因为库函数 gets 在复制字符时未对复制的字符数量进行检查，而只是简单地从标准输入中读取文本直到换行符（0a）出现，然后在这些文本的后面加上 NULL（00），将其复制到给定的缓冲区中。因此，当读入的文本长度超过分配给相应变量的存储空间大小时，将导致缓冲区溢出。事实上，前面提到的 Morris 蠕虫也是利用与本例相同的 C 语言标准库函数 gets 存在的缓冲区溢出漏洞展开攻击的。

```
int main(void ) {
    int IsSame=FALSE;
    char aa[8];
    char bb[8];
    aa=get_aa(char *label,"A",6);    //调用函数 get_aa 给 aa 赋值
    gets(bb);       //调用库函数 gets，从标准输入中读取字符串赋给 bb
    if(strncmp(aa, bb, 8)==0)
    IsSame =TRUE;
    printf("The string aa:%s and bb:%s Is the Same \n", aa, bb);
}
```

图 1-2　存在缓冲区溢出漏洞的 C 语言程序

假设调用函数 get_aa 后，变量 aa 的值为"AAAAAA"。当执行 gets 函数时，若从键盘上输入"BBBBBBBBBBBBBBBB"，则发生缓冲区溢出，gets 函数调用前后栈缓冲区的值如图 1-3 所示。由该图可知，变量 aa 的值已被破坏为"BBBBBBBB"。此时，程序的功能（比较变量 aa 和 bb 的值是否相同）显然已失效，因为只要输入任意一个前后对称的字符串（如本例中的 16 个 B），就必然使得变量 IsSame 的值为 1，而与变量 aa 原来的值无关。

内存地址	gets(bb) 调用之前	gets(bb) 调用之后	包含的值
...	...	...	
bffff01c	90900408	90900408	返回地址
bffff018	28f0ffbf (. . .	28f0ffbf (. . .	前帧指针
bffff014	00000000	01000000	IsSame
bffff010	08080808	00080808	
bffff00c	41410008 A A . .	42424242 B B B B	aa[4-7]
bffff008	41414141 A A A A	42424242 B B B B	aa[0-3]
bffff004	08080808	42424242 B B B B	bb[4-7]
bffff000	08080808	42424242 B B B B	bb[0-3]

图 1-3　缓冲区溢出导致数据遭到破坏

1.2.1.2 函数调用机制

当子函数被调用执行时,操作系统自动为之在栈区开辟一个属于该子函数的连续内存区域,该区域称为栈帧,栈帧的第一个内存单元称为栈底,而其最后一个内存单元称为栈顶。栈帧是一种数据结构,可方便地实现不同栈帧间的链接。在函数调用过程中,会涉及以下三个关键的寄存器:栈指针寄存器(Extended Stack Pointer,ESP)、基址指针寄存器(Extended Base Pointer,EBP)和指令指针寄存器(Extended Instruction Pointer,EIP)。栈指针寄存器,简称栈指针,该指针的值总是指向系统栈中最新一个栈帧的栈顶。基址指针寄存器,简称帧指针,该指针的值总是指向系统栈中最新一个栈帧的栈底。指令指针寄存器,简称指令指针,该指针永远指向内存指令区中下一条等待执行的指令的地址。

下面以图 1-2 所示的 C 语言程序的主函数 main 调用子函数 get_aa 来说明函数的调用机制。假设在函数 get_aa 中不再调用其他函数,调用函数 main 与被调用函数 get_aa 的栈帧结构如图 1-4 所示,函数调用过程主要包括以下四个阶段。

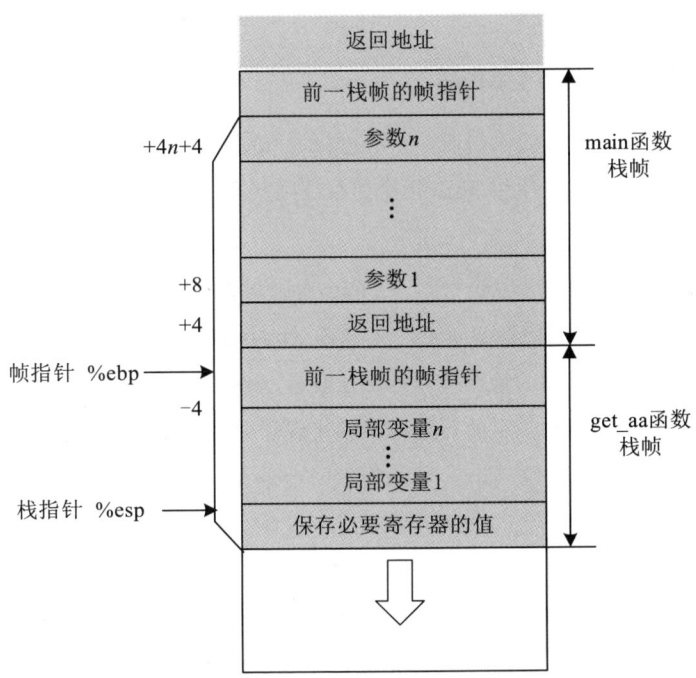

图 1-4 调用函数 main 与被调用函数 get_aa 的栈帧结构

1. 在调用函数 main 调用被调用函数 get_aa 时

- 在 main 的栈帧中,将 get_aa 的参数按声明位置逆序(从右向左)依次压入栈中。
- 把 main 函数当前指令的下一条指令地址(gets_aa 函数对应的指令地址,即返回地址)压入栈中。
- 使用 call 指令调用目标函数 get_aa。

2. 在被调用函数 get_aa 中

- 将当前帧指针 ebp 的值（主函数 main 栈帧的栈底单元地址）保存到栈中，即 push ebp。
- 将当前栈指针 esp 的值赋给帧指针 ebp，即 mov ebp, esp。显然，此时的栈指针和帧指针均指向同一个内存单元的地址，即函数 get_aa 栈帧的栈底地址。
- 为函数 get_aa 中的局部变量分配内存空间并改变栈顶，即 sub esp, MEM。这里用 MEM 表示局部变量所需要的存储空间（以字节为单位），其具体值在编译阶段由编译器根据 get_aa 函数中全部局部变量的大小确定。若被调用函数无局部变量，则跳过此步操作。
- 若需要保存某些寄存器的值，则将其压入栈，即 push REGS。这里用 REGS 表示在执行函数 get_aa 前需要保存的寄存器的值。
- 运行被调用函数 get_aa。在函数 get_aa 中，可以通过基址寻址方式访问主函数栈中的实参，如可通过地址%ebp+8 访问 get_aa 中的第 1 个实数，通过地址%ebp+12 访问 get_aa 中的第 2 个实数。

3. 在被调用函数 get_aa 运行结束后

- 保存被调用函数 get_aa 的返回值，一般可将返回值保存在寄存器 EAX 中。
- 若在调用函数 get_aa 前使用 push 指令保存了某些寄存器的值，则通过 pop 指令恢复相应寄存器的值。
- 将帧指针 ebp 的值赋给栈指针 esp，即 mov esp, ebp。这样，就释放了函数 get_aa 的局部变量占用的内存空间，同时恢复了前一个栈帧（main 栈帧）的栈顶。
- 将栈顶的值（main 栈帧的帧指针）弹出栈赋给 ebp，即 pop ebp。这样，就恢复了前一个栈帧（main 栈帧）的栈底。至此，通过栈帧实现了调用栈与被调用栈间的链接。
- 将控制权交还给调用函数 main，即执行返回指令 ret，将保存地址弹出栈。

4. 在调用函数 main 中

- 将被调用函数 get_aa 的参数从栈中弹出。
- 继续执行调用函数 main 中的下一条指令，即 gets(bb)语句对应的指令。

1.2.1.3 控制权非法转移

由前面的讨论可知，发生缓冲区溢出会破坏程序数据，在此基础上，本节将以实例说明缓冲区溢出是如何造成程序控制权非法转移、内存非法访问和程序异常终止等后果的。考虑图 1-2 所示的 C 语言程序，现在的目标是通过缓冲区溢出使得子函数 get_aa 运行结束后不返回调用它的主函数 main，而是重新再次执行自己。为此，需要事先找到装载 get_aa 函数的内存地址，以便用该地址改写栈中的返回地址。假设通过调试器已经找到了程序正常运行时函数参数、局部变量、返回地址、帧指针的内存分配及单元值，如表 1-1 所示。由该表可知，返回地址存储在主函数 main 栈帧地址为 bffff01c 的内存单元中，该单元存放的值是"08049090"，此值正是主函数 main 调用子函数 get_aa 结束后返回主函数继续执行的下一条指令对应的地址。

表 1-1　程序正常运行时的内存分配及单元值

内存区域	地址	单元含义说明（值）
栈区 （方向：自上而下）	bffff028 bffff020 bffff01c	主函数 main 栈底 参数 label 的地址（0804a23e） 返回地址（08049090）
	bffff018	get_aa 函数栈底（bffff028，前一栈帧的帧指针）
数据区	0805a124 0805a100	label（"buff"） （"erro"）
指令区 （方向：自下而上）	080490f0	…… 子函数 get_aa 首指令
	08049090	…… 下一条指令 call get_aa( ) …… 主函数 main

为了利用缓冲区溢出实现程序控制权的非法转移，首先必须确定填满缓冲区所需的字节数和关键内存单元的值，由图 1-3 可知，从分配给变量 bb 的首地址 bffff000 到位于帧指针地址下方单元的地址 bffff017，一共有 24 字节的存储空间，现选择字符串"ABCDEFGHIJKLMNOPQRSTUVWX"作为填充字符。帧指针保持原来的值不变：bffff028。返回地址由原来的 08049090（指向主函数调用子函数 get_aa 命令的下一条指令，即正常返回到主函数中）改写为 080490f0（指向子函数 get_aa 在指令区的首地址，即改变程序的控制权以便再次执行子函数）。由于帧指针和返回地址均采用小端存储，因此在从标准设备输入时应注意字节顺序。帧指针值应按顺序"28f0ffbf"输入，而返回地址的正确顺序为"f0900408"。综上所述，得到了从标准设备中输入的十六进制串"41424344 45464748 494a4b4c 4d4e4f50 51525354 55565758 28f0ffbf f0900408"。至此，利用缓冲区溢出实现了程序控制权的非法转移，其结果如图 1-5 所示。

内存地址	gets函数被调用前	gets函数被调用后	单元含义
...	...	...	
bffff020	24a10508 $...	00a10508	label
bffff01c	90900408	f0900408	返回地址
bffff018	28f0ffbf (...	28f0ffbf (...	前帧指针
bffff014	00000000	55565758 UVWX	IsSame
bffff010	08080808	51525354 QRST	
bffff00c	41410008 AA..	4d4e4f50 MNOP	aa[4-7]
bffff008	41414141 AAAA	494a4b4c IJKL	aa[0-3]
bffff004	08080808	45464748 EFGH	bb[4-7]
bffff000	08080808	41424344 ABCD	bb[0-3]

图 1-5 缓冲区溢出导致控制权非法转移

1.2.2 拒绝服务攻击

拒绝服务（Denial of Service，DoS）攻击的本质是对可用性的破坏，攻击者通过耗尽目标机器的资源致使其无法向用户提供服务，是黑客常用的攻击手段之一。攻击者往往可以通过消耗目标机器的网络带宽、CPU、内存、磁盘空间等系统资源来达到拒绝服务攻击的目的。例如，攻击者进行拒绝服务攻击时，通过填满服务器的缓冲区，使之不能接收新的请求；或者通过 IP 地址欺骗，迫使服务器把合法用户的连接复位，影响合法用户的连接。拒绝服务攻击问题一直得不到合理有效的解决是多方面原因造成的，其中一个重要的原因是网络协议本身存在安全缺陷，因而拒绝服务攻击也成为攻击者的常用攻击手法。下面对 SYN 泛洪攻击、源地址欺骗、UDP 泛洪攻击、死亡之 Ping、泪滴攻击、LAND 攻击、蓝精灵攻击和 Fraggle 攻击等 8 种典型的拒绝服务攻击进行分析讨论。

1. SYN 泛洪攻击

SYN 泛洪攻击（SYN flood attack）是当前最流行的拒绝服务攻击之一，该攻击利用 TCP 协议缺陷发送大量伪造的 TCP 连接请求，以耗尽对方的内存、CPU 等系统资源。

SYN 泛洪攻击利用了 TCP 协议在建立连接时使用的三次握手（three way handshake）机制。假设一个客户端正处于 TCP 连接阶段，在完成第一次握手（向服务器发送了 SYN 报文）后突然死机或掉线，则服务器在发出 SYN+ACK 应答报文（第二次握手）后是无法收到第三次握手的 ACK 报文的，因此就无法完成 TCP 连接。为了应对这种意外情况，TPC 协议规定服务器应该在超时后再次发送 SYN+ACK 报文给客户端，若在等待一段时间（通常称为"TCP 同步超时"）后依然无法收到期待的 ACK 报文，则丢弃这个未完成的半连接。TCP 同步超时的时长一般为 30~120s，若半连接的数量很多，则会给服务器带来两个方面的后果：一方面，服务器为了维护大量这样的半连接而不得不消耗相当多的内存资源，因为每个半连接列表都会占用一定的缓存空间；另一方面，随着半连接列表数量的增多，服务器定期遍历这些表项也会消耗不少的 CPU 时间。因此，当服务器的 CPU、内存等资源不够充足时很容易受到 SYN 泛洪攻击。

2. 源地址欺骗

源地址欺骗（source address spoofing）是指攻击者在发送攻击报文时使用了虚假的源地址。源地址欺骗一般很少单独使用，通常伴随其他类型攻击共同出现，比如利用 TCP 首部的 RST 位来实现对目标主机的攻击。假设有一个源地址为 A 的合法用户已经同服务器建立了正常的连接，攻击者构造攻击报文时，在将源地址伪装为 A 的同时置位 RST 位，然后向服务器发送该攻击报文。服务器接收到该报文后，便认为自己与 A 的连接有误，于是清空缓冲区中建立好的连接。此后，若源地址为 A 的合法用户再发送数据，由于服务器已经没有对应的连接了，因此它就必须重新开始建立连接。在实际攻击时，攻击者会伪造大量的 IP 地址，向目标发送大量的 RST 报文，使服务器不能对合法用户提供服务，从而实现对受害服务器的拒绝服务攻击。

3. UDP 泛洪攻击

UDP 泛洪攻击（UDP flood attack）是指攻击者通过向目标系统的开放 UDP 端口发送大量的 UDP 报文，以耗尽对方的网络带宽资源，进而达到拒绝服务攻击的目的。比如，攻击者可以借助 Chargen、Echo 等 UDP 服务来实施攻击：Chargen 服务器收到客户端请求后，将向客户端发送最大长度为 512 字节的随机字符串，Chargen 的默认端口号为 19；而 Echo 服务器在收到客户端请求后会发回和请求报文内容相同的 UDP 应答报文，默认端口号为 7。攻击者在构造 UDP 攻击报文时，将源地址伪造成某台提供 Echo 服务的服务器地址，同时将源端口号和目的端口号分别设置为 7 和 19。攻击者向 Chargen 服务器发送该 UDP 攻击报文，由于源地址已修改为 Echo 服务器，Chargen 服务器将向 Echo 服务器的 7 号端口发送 Chargen 应答报文；Echo 服务器收到 Chargen 报文后，将向 Chargen 服务器发回 Echo 应答报文。这样，就在这两台服务器之间生成大量的数据流，这些数据将会消耗大量的网络带宽并最终导致拒绝服务攻击。

4. 死亡之 Ping

死亡之 Ping（Ping of death）的基本思想是通过向目标发送巨型分组使对方在重组分片时发生缓冲区溢出，从而导致死机。IP 协议规定分组的最大长度为 64KB，而不同数据链路层的最大传输单元 MTU 的值是不一样的，当超过 MTU 允许的大小时就需要分片，分片操作可以由源主机或者中间路由器来完成。分片后的各个小分片需要重组成一个分组，在重组时若不检查缓冲区的大小，则会出现内存分配错误，导致 TCP/IP 协议栈崩溃，最终致使接收方死机。

5. 泪滴攻击

泪滴攻击（teardrop）的基本原理与死亡之 Ping 类似，也是利用 IP 协议在重组时存在漏洞实施攻击的。不同的是，泪滴攻击者在构造攻击分片时，有意设置片偏移量使得在不同的分片中存在相同的编号。受攻击的主机在接收到这些分片并进行重组时，若不对偏移量进行检查，由于分片间存在重叠部分，重组无法完成，从而导致 IP 协议崩溃。

6. LAND 攻击

LAND 攻击（Local Area Network Denial attack）是一种局域网拒绝服务攻击，其攻击原理是攻击者在构造攻击 TCP 连接的 SYN 报文时，将源 IP 地址和目的 IP 地址都设置成受害主机的 IP 地址。受害主机收到此 SYN 报文后，将向自己发送 SYN-ACK 报文，之后又发回 ACK 报文完成三次握手并创建一个空连接。被攻击的主机每接收到一个 LAND 攻击报文，都将建立并维护这样的空连接直到超时。一般来说，不同操作系统的服务器对 LAND 攻击的抵御能力不同：UNIX 服务器往往很快崩溃；而 Windows 服务器会变得极其缓慢，大约持续 5min 后才会崩溃。

7. 蓝精灵攻击

蓝精灵攻击（Smurf attack）也是一种面向局域网的拒绝服务攻击，其攻击原理是攻击者在构造类型为"Echo request"的 ICMP 攻击报文时，将源 IP 地址伪造成受害主机的 IP 地址，而将目的 IP 地址设置为受害主机所在网络的子网广播地址。攻击者发出蓝精灵攻击报文后，受害主机所在子网的所有主机均收到此 ICMP 攻击报文，每台主机均会向受害主机发回一个类型为"Echo reply"的 ICMP 应答报文。可见，受害主机所在子网的主机的数量越多，ICMP 应答报文就越多，攻击效果也越明显，受害主机将越快崩溃。一般来说，蓝精灵攻击比死亡之 Ping 的攻击流量可以高出 1~2 个数量级。

8. Fraggle 攻击

从封装层次上看，Fraggle 攻击是蓝精灵攻击和 UDP 泛洪攻击的结合体：Fraggle 攻击和蓝精灵攻击在网络层是一样的，而 Fraggle 攻击和 UDP 泛洪攻击在运输层是一样的。Fraggle 攻击的基本原理是攻击者在构造 UDP 攻击报文时，将源 IP 地址伪造成受害主机的 IP 地址，而将目的 IP 地址设置为受害主机所在网络的子网广播地址；同时，将源端口号设置成 7（Echo 服务的默认端口号），目的端口号设置为 19（Chargen 服务的默认端口号）。当受害主机子网内的所有主机都开放端口 19，而受害主机开放端口 7 时，攻击效果将达到最强。若相应端口未开放，则会在网络中产生大量诸如"目的端口不可达"的 ICMP 报文。

1.2.3 网络犯罪的特点

同传统的犯罪相比，网络犯罪具有以下特点。

1. 成本低，传播迅速，传播范围广

发送电子邮件比传统寄信所花的成本少得多，尤其是发送到国外的邮件。随着网络的发展，只要敲一下键盘，几秒钟就可以把电子邮件发给众多的人。理论上，接收者可以是全世界的人。

2. 互动性、隐蔽性强，取证困难

网络发展形成了一个虚拟的计算机空间，既消除了国境线，也打破了社会和空间界限，使得双向性、多向性交流传播成为可能。在这个虚拟空间里，对所有事物的描述都仅仅是一堆冷冰冰的密码数据，因此谁掌握了密码就等于获得了对财产等事物的控制权，就可以在任何地方登录网站。

3. 社会危害性严重

随着计算机信息技术的不断发展，从国防、电力到银行和电话系统均实现了数字化、网络化，一旦这些部门遭到侵入和破坏，后果将不堪设想。

4. 网络犯罪是典型的计算机犯罪

网络犯罪中比较常见的偷窥、复制、更改或者删除计算机数据、信息的犯罪，以及散布破坏性病毒、逻辑炸弹或者放置后门程序的犯罪，都是典型的以计算机为对象的犯罪；而网络色情传播犯罪，网络侮辱、诽谤与恐吓犯罪，以及网络诈骗、教唆等犯罪，则是以计算机网络形成的虚拟空间作为犯罪工具、犯罪场所进行的犯罪。

1.2.4 网络犯罪者

谁是网络犯罪者？显而易见，他们是网络空间信息的普通使用者。随着用户的膨胀，其中的罪犯数量也会快速增加。很多研究表明，下列群体很可能就是网络犯罪之源。

1. 黑客

黑客实际上是了解很多计算机和计算机网络知识并将这种知识用于犯罪目的的计算机爱好者。随着 20 世纪 80 年代以来互联网的普及，计算机黑客的数量一直处于有增无减的状态。

2. 犯罪团伙

不同的网络犯罪往往具有不同的犯罪动机。例如，具有黑客能力的犯罪团伙入侵信用卡公司的系统盗窃数千张信用卡的账号。

3. 经济间谍

网络空间和电子商务的增长和全球化引发了更多的网络犯罪，有组织的经济间谍在网络上搜寻各大公司的秘密。随着原发性研发成本的飙升以及市场经济全球化带来的市场竞争，有些公司试图通过网络盗窃商业、市场和其他公司秘密，以进一步牟取暴利。

4. 心怀不满的前雇员

许多研究显示，有些因为劳资纠纷而被迫离开的前雇员对之前工作的企业心怀不满，为此经常将前雇主作为攻击对象，由此而引发进一步的网络犯罪。通常，前雇员对前雇主的机密信息都有些许接触，这在一定程度上给随之到来的网络犯罪带来了便利。

5. 内部人员

心怀不满的内部人员可能是网络犯罪的主要来源，因为他们不需要了解很多有关受害者计算机系统的知识。在许多情况下，他们每天都在使用系统，这为他们任何时候不受限制地访问系统提供了便利。1999 年，美国计算机安全协会（Computer Security Institute，CSI）与联邦调查局（Federal Bureau of Investigation，FBI）联合进行的调查显示，有 55%的受访者报告恶意行为是内部人员所为。因此，对于内部人员，特别是掌握核心机密的内部人员，公司必须给予足够的重视和监督，否则将给公司带来难以想象的损害。

1.2.5 黑客拓扑结构

前面已经指出，黑客通常是非常了解计算机和计算机网络工作原理的计算机爱好者，他们运用所学知识策划针对系统的攻击。成熟的黑客会预先策划好攻击但不会影响未标记的系统成员。

为了达到这种精度，通常需要使用指定的拓扑攻击模式。依据这些拓扑，黑客就可以在众多的网络主机中选择目标受害主机。具体的攻击模式、拓扑会受到以下因素和网络配置的影响。

> 设备的可用性。如果受害者仅是一台主机，这就会显得尤为重要。攻击的时候必须保证仅针对这一台主机而不会影响其他主机，否则攻击就不能进行。
> 互联网接入可用性。如前所述，选择的受害者主机或网络必须是可达的。
> 网络环境。攻击的时候往往要根据受害主机或子网所在的环境，小心隔离目标单位以便不会影响其他主机。
> 安全范围。在针对系统进行攻击时，黑客往往会事先确定攻击的安全范围，以便攻击时不会被发现。

综上所述，选择的攻击模式主要是根据受害者类型、位置、分发方法等确定的，主要存在以下四种模式。

1. 一对一

这种攻击由某个攻击者发起，针对一个已知的受害者。这种攻击是已知的攻击，攻击者知道甚至熟悉受害者，有时受害者也可能知道攻击者。其拓扑结构如图1-6所示。

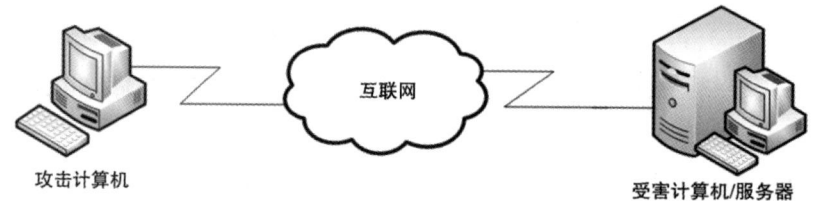

图 1-6　一对一攻击模式的拓扑结构

2. 一对多

这种攻击是匿名的，大多数情况下，攻击者不知道受害者是谁，他们对受害者来讲也是匿名的。这种攻击在最近几年兴起，是最容易实现的攻击模式之一，其拓扑结构如图1-7所示。

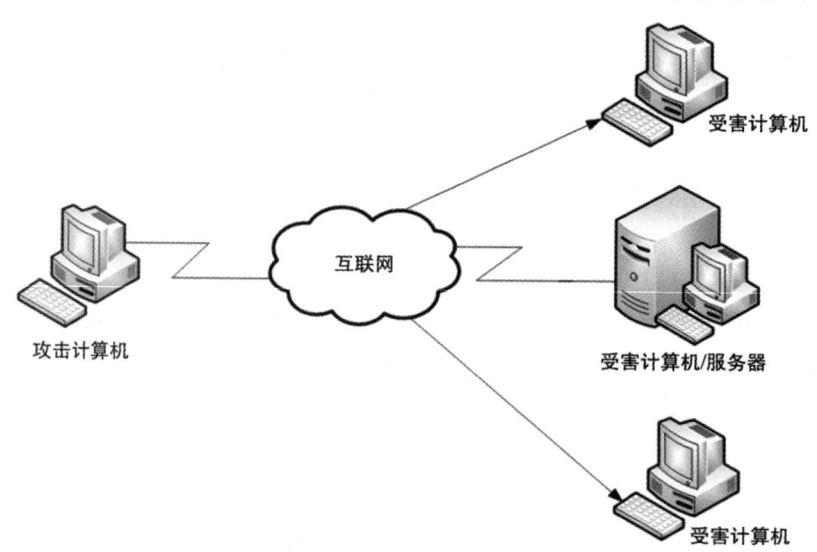

图 1-7　一对多攻击模式的拓扑结构

3. 多对一

到目前为止，这种攻击还很少，但随着分布式拒绝服务（DDoS）攻击在黑客团体中越来越受宠，这种攻击有增多的迹象。在这种攻击中，攻击者通过一台主机欺骗其他主机（二手受害者），然后将其用作对最终受害者进行雪崩效应攻击的新的来源。其拓扑结构如图 1-8 所示。

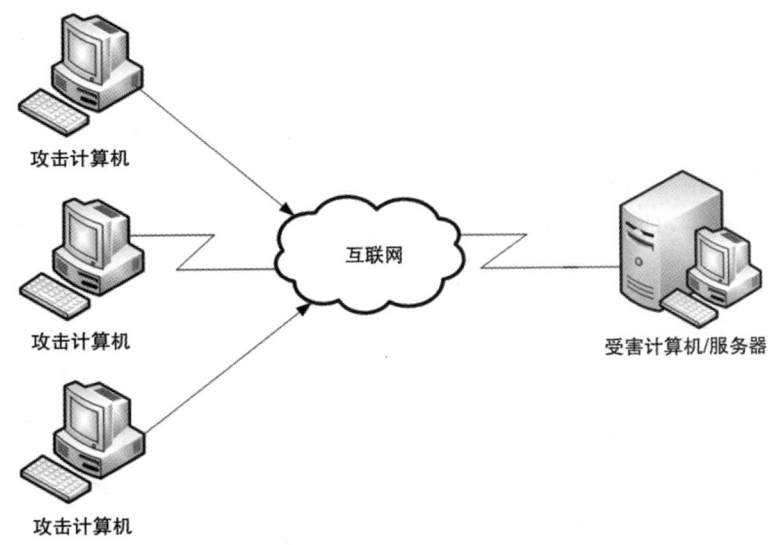

图 1-8　多对一攻击模式的拓扑结构

4. 多对多

这种攻击很少，但最近的有关报告指出其有所增多。例如，在某些 DDoS 攻击案例中，就有被攻击者选作二手受害者的一组站点的情况，然后这些站点常常被用来"攻击"所选择的受害组。每一组中涉及的数目变化都很大，可以从几个到几千个。就像多对一攻击一样，攻击者采用这种攻击模式需要很好地理解网络基础设施，精心选择二手受害者和最终受害者群体。其拓扑结构如图 1-9 所示。

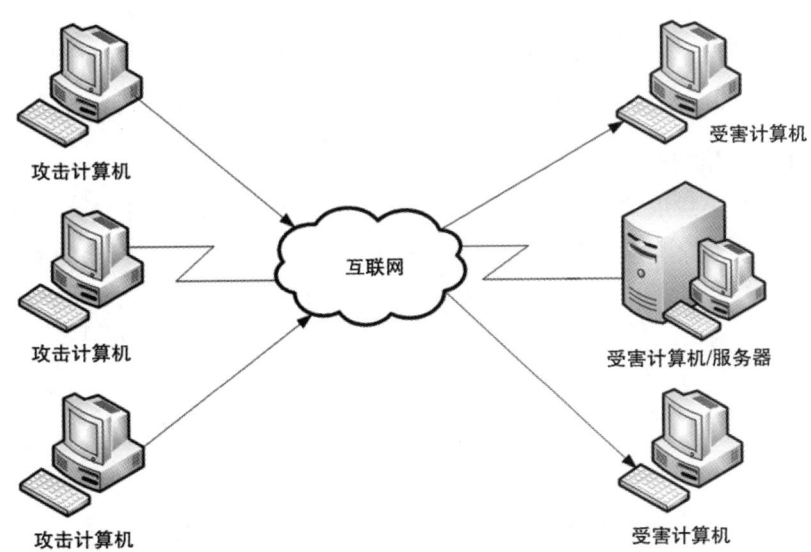

图 1-9　多对多攻击模式的拓扑结构

1.2.6 黑客攻击发展趋势

黑客攻击发展趋势具有以下 7 个方面的特点。

1. 手段高明化

黑客界已经意识到单靠一个人的力量远远不够了，逐步形成了一个团体，利用网络进行交流和团体攻击。

2. 活动频繁化

黑客不再需要掌握大量的计算机和网络知识，学会使用几个黑客工具，就可以在互联网上进行攻击活动。黑客工具的大众化是黑客活动频繁的主要原因。

3. 动机复杂化

黑客攻击的动机目前已经不再局限于为了国家、金钱和刺激，而是和国际政治变化、经济变化紧密地结合在一起。

4. 黑客年轻化

基于互联网的普及，很多中小学生也可以接触到世界各地形形色色的信息资源，所以黑客朝着年轻化的方向发展。

5. 黑客的破坏力增强

随着互联网的普及、电子商务的蓬勃发展，社会对互联网的依赖性日益增加，网络黑客的破坏力也随之增强。仅在美国，黑客每年造成的经济损失就超过 100 亿美元。可想而知，每年全球因为黑客攻击遭受的损失有多大。

6. 黑客技术普及化

黑客组织的形成和傻瓜式攻击工具的大量出现导致的一个直接后果就是黑客技术普及。在互联网上，可供黑客之间交流的站点比比皆是，随意搜索一下就能找到一大堆。这些黑客站点上面提供黑客攻击工具，公布系统漏洞，公开传授黑客攻击技术，发布各种黑客知识、系统软件源码，这些因素在很大程度上推动了黑客技术的普及，吸引着更多的人参与其中。

7. 黑客组织化、团体化

随着人们网络安全意识的增强、计算机产品安全性能的提高，软件系统漏洞越来越难被发现，单个黑客想要对某个攻击目标造成破坏也变得越来越困难。受利益的驱使，曾经的单一黑客开始寻找"盟友"，组团作战。群体性的黑客攻击往往成功率更高，对目标造成的伤害也更大。相应地，黑客还能从中学到一些新的攻击技术。以上种种因素导致了黑客的组织化、团体化。

本章小结

本章主要介绍了网络犯罪与黑客的基本知识。其中，针对网络犯罪，从其基本概念、实施的方法、特点等几个方面进行了阐述；对于黑客，则从其发展历史、分类、攻击拓扑结构、常用的攻击工具与攻击手段、攻击步骤及发展趋势等几个方面进行描述。希望通过本章的学习，读者能够对网络犯罪的概念、实施网络犯罪的方法、网络犯罪的特点、常见的拒绝服务攻击类型以及黑客的基本定义、类型、常用的系统攻击工具、常用的攻击手段、攻击步骤等有一定的了解。此外，有兴趣的读者还可以自行查阅有关资料，了解一些常用的网络犯罪应对措施。

第 2 章 密码学

我们在生活中有许多隐私不想让其他人知道，更不想让其他人广泛传播或者使用。对于我们来说，这些隐私是至关重要的，它们是我们个人的重要信息，其他人不需要知道。为了防止隐私泄露，我们需要设置密码，从而保护信息安全。那么密码到底是干什么的呢？其实，密码就是用来防止未被允许的陌生人进入你的"账户""系统"等读写你的文件和数据的。通信设备中使用密码，个人在银行取款时使用密码，登录计算机时使用密码，开启保险箱时使用密码，儿童玩电子游戏时也使用密码……密码在我们生活中无处不在。不过，这些密码只是一种特定的暗号或口令。现代的密码已经比古代有了长远的发展，并逐渐形成一门科学。从专业上来讲，密码是按特定法则编成，用于通信双方的信息明密变换的符号。研究密码的学科称为密码学。密码主要用于保护传输和存储的信息；除此之外，密码还用于保证信息的完整性、真实性、可控性和不可否认性。密码学是信息安全技术的核心基础，主要由密码编码技术和密码破译技术两个分支组成。在密码学研究发展的过程中，密码编码者一直努力分析密码算法的特性，试图证明其安全性；与此同时，另一部分人则同样对密码算法进行分析，但是以破译为目的。作为密码学的两个方面，密码编码与密码破译这对孪生兄弟始终如影随形，正是这种对立统一关系推动了密码学自身的发展。

2.1 基础知识

在密码学中，有一个五元组：{明文，密文，密钥，加密算法，解密算法}，对应的加密方案称为密码体制。

明文：输入的原始信息，即消息的原始形式，通常用 m 或 p 表示。所有可能明文的有限集称为明文空间，通常用 M 或 P 表示。

密文：明文经加密后的结果，即消息被加密处理后的形式，通常用 c 表示。所有可能密文的有限集称为密文空间，通常用 C 表示。

密钥：参与密码变换的参数，通常用 k 表示。一切可能的密钥构成的有限集称为密钥空间，通常用 K 表示。

加密算法：将明文变换为密文的变换函数，相应的变换过程称为加密，即编码的过程〔通常用 E 表示，即 $c=Ek(p)$〕。

解密算法：将密文恢复为明文的变换函数，相应的变换过程称为解密，即解码的过程〔通常用 D 表示，即 $p=Dk(c)$〕。

有实用意义的密码体制要满足 $p=D_k(E_k(p))$，即用加密算法得到的密文总能用一定的解密算法恢复出原始的明文。而密文的获取同时依赖于明文和密钥。

2.1.1 可能的攻击

根据攻击者对明文、密文等信息掌握的多少，可将攻击分为以下几种情形。

唯密文攻击（ciphertext only attack）：攻击者知道密码算法，但仅能根据截获的密文进行分析，以得出明文或密钥。攻击者所能利用的数据资源仅为密文，这是对他最不利的情况。

已知明文攻击（known plaintext attack）：攻击者除有截获的密文外，还有一些已知的"明文—密文对"来破译密码。攻击者的任务目标是推出用来加密的密钥或某种算法，这种算法可以对用该密钥加密的任何新的消息进行解密。例如，假设攻击者截获了一份加密的通信稿，并在第二天看到了解密后的通信稿。如果加密者没有改变密钥而攻击者又能推断出解密密钥，那么攻击者就能读出将来的所有信息。或者，如果加密者总是以某些的固定字符开头，那么攻击者就有了一小份密文和相对应的明文。对许多弱密码系统来说，这足以找到密钥。

选择明文攻击（chosen plaintext attack）：攻击者有机会接触加密机，他不能打开加密机找密钥，但可以加密大量经过适当选择的明文，然后试着利用所得的密文来推测密钥。

选择密文攻击（chosen ciphertext attack）：攻击者有机会接触加密机，用它对若干字符串进行解密，然后试着用所得结果推断密钥。

自适应选择明文攻击（adaptive-chosen-plaintext attack）：这是选择明文攻击的一种特殊情况，指的是攻击者不仅能够选择要加密的明文，还能够根据加密的结果对以前的选择进行修改。

选择密钥攻击（chosen-key attack）：这种攻击情况在实际应用中比较少见，它仅表示攻击者了解不同密钥之间的关系，并不表示他能够选择密钥。

选择文本攻击（chosen text attack）：这种攻击情况是选择明文攻击与选择密文攻击的结合。攻击者已知的东西包括加密算法、由密码破译者选择的明文消息和它对应的密文，以及由攻击者选择的猜测性密文及其对应的已破译的明文。

现代密码学中一个最重要的假设就是 Kerckhoffs 原则：在评估一个密码系统的安全性时，必须假定敌方知道所用的加密方法。敌方有很多方式可以获得这个信息，如加密机被俘获和分析、人员被捕或叛变，因而密码系统的安全性应该基于密钥而不是所用算法的隐蔽性。我们总假定攻击者知道用来加密的算法。

2.1.2 单向函数与单向 Hash 函数

1. 单向函数

单向函数是正向计算起来相对容易但逆向计算异常困难的函数。也就是说，若已知 x，则很容易计算出 $f(x)$；若已知 $f(x)$，却很难计算出 x。注意，这里的"难"指的是在计算上是不可行的，比如即使将世界上所有的计算机都用来计算，从 $f(x)$ 计算出 x 也要花费相当长（如数百万年）的时间。

陷门单向函数是有秘密陷门的特殊单向函数，它在一个方向上易于计算而在反方向上难以计算；但是，如果你知道那个秘密，则很容易在另一个方向上计算这个函数。也就是说，已知 x，易于计算 $f(x)$，而已知 $f(x)$，却难以计算 x；然而，存在秘密信息 y，一旦给出 $f(x)$ 和 y，就很

容易计算 x。

2. 单向 Hash 函数

单向 Hash（哈希）函数有很多名字：压缩函数、缩短函数、消息摘要、指纹、密码校验和、信息完整性检验（DIC）、操作检验码（MDC）等。

Hash 函数就是把可变长度输入串转换成固定长度（经常更短）输出串（叫作 Hash 值）的一种函数。

Hash 函数是公开的，对处理过程不保密。单向 Hash 函数的安全性在于它的单向性。平均而言，预映射的值的单个位的改变，将引起 Hash 值中一半位的改变。已知一个 Hash 值，要找到预映射的值使它的 Hash 值等于已知的 Hash 值在计算上是不可行的。

2.1.3 对称密码和非对称密码

网络安全通信中要用到两类密码算法：一类是对称密码算法，另一类是非对称密码算法。对称密码算法有时又叫传统密码算法、秘密密钥算法或单密钥算法，非对称密码算法也叫公开密钥密码算法或双密钥算法。对称密码算法的加密密钥能够从解密密钥中推算出来，反过来也成立。在大多数对称密码算法中，加密和解密密钥是相同的。它要求发送者和接收者在安全通信之前，商定一个密钥。对称密码算法的安全性依赖于密钥，泄漏密钥就意味着任何人都能对消息进行加密/解密。只要通信需要保密，密钥就必须保密。

对称密码算法又可分为两类：一类是一次只对明文中的单个位（或字节）运算的算法，称为序列算法或序列密码；另一类是对明文的一组位进行运算的算法，这些位组称为分组，相应的算法称为分组算法或分组密码。现代计算机密码算法的典型分组长度为 64 位——这个长度既考虑到分析破译密码的难度，又考虑到使用的方便性。后来，随着破译能力的发展，分组长度提高到了 128 位或更多。

采用对称密码算法的主要问题是密钥的生成、注入、存储、管理、分发等很复杂，特别是随着用户的增加，密钥的需求量成倍增加。在网络通信中，大量密钥的分配是一个难以解决的问题。例如，若系统中有 n 个用户，其中每两个用户之间需要建立密码通信，则系统中每个用户须掌握 $n-1$ 个密钥，而系统中所需的密钥总数为 $n \times (n-1)/2$ 个。对于有 10 个用户的情况，每个用户必须掌握 9 个密钥，系统中密钥的总数为 45 个。对于有 100 个用户的情况，每个用户必须掌握 99 个密钥，系统中密钥的总数为 4950 个。这还是用户之间的通信只使用一种会话密钥的情况。如此数量庞大的密钥的生成、管理、分发确实是一个难处理的问题。

1976 年，美国斯坦福大学的学者迪菲和赫尔曼提出了非对称密码算法。所谓非对称密码算法，就是指使用不同的加密密钥与解密密钥，是一种"由已知加密密钥推导出解密密钥在计算上不可行"的密码算法。

在非对称密码算法中，加密密钥（公开密钥）PK 是公开信息，而解密密钥（私人密钥）SK 是需要保密的。加密算法 E 和解密算法 D 也都是公开的。虽然私人密钥 SK 是由公开密钥 PK 决定的，但不能根据 PK 计算出 SK。

与对称密码算法不同，该算法采用两个不同的密钥对信息进行加密和解密。每个用户的加密算法 E 和解密算法 D 须满足以下条件：

> D 是 E 的逆，即 $D[E(x)]=x$；
> E 和 D 都容易计算；
> 由 E 求解 D 十分困难。

从上述条件可以看出，在非对称密码算法中，加密密钥不等于解密密钥。加密密钥对外公开，任何用户都可用加密密钥将传送给他的信息加密发送，而其唯一保存的解密密钥是保密的，也只有它能将密文复原、解密。虽然解密密钥理论上可由加密密钥推算出来，但实际上是不可能的；或者虽然能够推算出，但要花费很长的时间，从而不可行的。所以，将加密密钥公开也不会危害密钥的安全。

2.1.4 密钥长度

密码算法的安全性是一个很难度量的性质。大多数算法会使用密钥，算法的安全性与敌方确定密钥的困难程度有关。最直接的方法是尝试每一个可能的密钥，看哪些能得到有意义的密文，这种方法被称为蛮力攻击（brute force attack）。在蛮力攻击中，密钥的长度和搜索整个密钥空间所需的时间直接相关。例如，一个密钥的长度是 16 位，那么就有 $2^{16}=65536$ 个可能的密钥。DES 算法的密钥长度为 56 位，从而有 $2^{56}\approx7.2\times10^{16}$ 个可能的密钥。这么看来，一个密码系统似乎可以通过简单地尝试所有可能的密钥而被破解，但事实并非如此。假定需要尝试 10^{30} 种可能性，而你的计算机每秒可做 10^9 次这种运算，一年大约有 3×10^7 秒，所以需要大概 3×10^{13} 年才能完成这个任务，这比所预测的宇宙寿命还长。

较长的密钥更有优势，但并不保证会增加敌方破译密码的困难性，算法本身也能起到关键性的作用。有些算法可能被蛮力之外的其他方法所攻击，还有些算法并不能很有效地使用密钥位。有一点非常重要，需要牢记：不能对所有的 128 位算法一视同仁。

例如，一个最容易破解的密码系统就是替换密码，虽然这个密码的可能密钥数为 $26!\approx4\times10^{26}$ 个。相比较而言，复杂的 DES 算法有 $2^{56}\approx7.2\times10^{16}$ 个可能的密钥，在一个特别设计的计算机上找一个 DES 密钥通常需要一天的时间。两者的区别在于：对替换密码的攻击利用的是语言的基本结构；而对 DES 算法的攻击是蛮力攻击，要尝试所有可能的密钥。

蛮力攻击应该是最后使用的手段，密码分析员总是希望能找到比这更快的攻击方法。比如，对于替换密码可以用频率分析，对于离散对数可以用生日攻击等方法。

注意，一个算法现在看起来安全，并不意味着以后也安全。人类已经开发出了各种具有创造性的密码攻击方法。现代密码学中有很多算法或协议被成功攻破的例子，原因有很多，或者是系统内在的漏洞，或者是技术的进步，或者是人类对数学研究的加深。DES 算法经受住了 20 年的考验，最终被一台精心设计的并行计算机攻破。当你在读这本书时，量子计算的研究正在进行，它将极大地改变密码算法的面貌。

例如，有一个依赖 200 位大整数因子分解的困难性的密码系统，假定分解这般大小的一个数 n，如果方法是用 n 除以所有不超过 n 的平方根的素数，那么对于小于 10^{100} 的素数是不可能的，因为其素数个数大概是 4×10^{97}！显然，必须使用更精致的因子分解算法，而不是蛮力攻击。当 RSA 算法被发明时已经有一些好的因子分解算法可用，但是据当时估计，要分解一个 129 位数，在可预见的很长一段时间里是不可能的。然而算法和计算机体系结构的进步已经使得那样的分解成为很平常的事（虽然仍需要大量的计算资源），所以为了安全性，现在通常推荐使用几百位的整数。但对于一台量子计算机而言，分解这些数也轻而易举，因而整个 RSA 算法（以及其他许多方法）都需要重新考虑。

因此，我们自然会产生两个疑问：是否有不可破译的密码系统？为什么不能总是用这种系统？

第一个疑问的答案是肯定的，有一个被称为一次一密（one-time pad）的系统就是不可破译的，甚至用蛮力攻击也得不到密钥。但不幸的是，使用一次一密的代价非常昂贵，它需要交换一个明文和一个一样长的密钥，而这个密钥只能用一次。因此，人们会选择当操作正确时，密钥长度适当、在合理时间内不被攻破的算法。

在考虑密钥大小时很重要的一点是：在大多数情况下，虽然增加密钥长度可以在数学上提高安全性，但在实际中并不总是可行的。如果你用的芯片字长是 64 位的，那么密钥长度从 64 位增加到 65 位就可能意味着硬件要重新设计，而这个代价非常高。因此，设计好的密码系统需要同时考虑数学和工程两个方面。

最后，我们讨论一下数的大小。有两种方法可以衡量数的大小：数 n 的实际大小和数的十进制表示的位数（也可以用它的二进制表示），后者大约是 $\lg(n)$。用标准算法算一个 k 位数 n 的平方所需做的个位数乘法的数目是 k^2，或大约是 $\lg(n)^2$。为分解数 n，用所有不超过 n 的平方根的素数去除 n 所要做的除法次数大约是 $n^{1/2}$。一个运行时间为 $\lg(n)$ 的幂的算法比运行时间为 n 的幂的算法要可取得多。在这个例子中，如果让 n 的位数翻一倍，那么计算 n 的平方所需的时间就会增加到 4 倍，分解 n 所需的时间会剧增。当然，有更好的算法可用来进行这两种运算，但就目前来讲，分解运算所需的时间远远超过乘法运算的时间。

2.1.5 密码学应用

密码学不仅关乎信息的加密和解密，也涉及解决现实世界中需要信息安全的问题，其主要目标有四个。

- ➢ 机密性。攻击者不能读到正常通信双方发送的消息，其主要工具是加密算法和解密算法。
- ➢ 数据完整性。接收方想确定接收到的消息没有被更改。例如，可能会发生传输错误，或者敌方可能窃听了传输过程并在消息到达预定的接收方之前进行了篡改。许多密码学原语（如 Hash 函数）都提供了检测数据是否受到敌方恶意或突然操控的方法。
- ➢ 鉴别。接收方想确定他所接收到的消息是特定的人发送的。实际上密码学中有两种类型的鉴别：实体鉴别和数据源鉴别。"身份识别"这个术语通常用来说明验证通信各方身份的实体鉴别。数据源鉴别则是把诸如数据的创建者和创建时间之类的原始信息与数据绑定在一起。
- ➢ 不可抵赖性。发送方不能声称他没有发送消息。例如，在电子商务应用中，消费者不能否认自己已做出的认购，这点很重要。

鉴别和不可抵赖性是紧密关联的概念，但也有一个区别。在对称密钥密码系统中，接收方（如张三）可以确定消息的确来自某个发送者（如李四，假设密钥只有张三和李四知道），因为其他人不能对张三成功解密的消息进行加密，因此鉴别是自动的。但是张三不能向别人证明该消息是李四发送的，因为也有可能是张三自己发送的消息。所以，不可抵赖性在这种系统中本质上是不可能实现的。但在，在公钥密码系统中，鉴别和不可抵赖性都能实现。

基于此，密码学的具体应用大体上有以下几种。

➢ 数字签名。纸质文件的一个最重要的特征就是它的签名。在一份文件署名后，个体的身份就和消息结合在一起了，这使得其他人要在另一份文件上伪造签名是很困难的。但电子信息很容易被完全复制，怎么才能防止敌方把一份文件的签名剪贴到另一份电子文件上呢？通过对电子信息进行签名的密码协议可以实现，它使得每个人都相信电子信息的签名者就是文件的签名人，并且签名人不能否认对文件的签署。

➢ 身份识别。当登录一台机器或创建一个通信链接时，用户要验证自己的身份。但简单地输入用户名是不够的，因为这不能证明这个用户就是他（她）所声称的那个用户，通常还要使用密码。我们将会接触各种验证身份的方法。

➢ 秘密分享。假设你有一个银行保险箱的密码，但你不想把这个密码托付给单个人，而是想把它分享一群人，这样要打开保险箱就至少要他们当中的两个人在场。秘密分享就用于解决这个问题。

➢ 安全协议。怎样才能在像互联网这样的公开渠道上安全交易？怎么保护信用卡上的信息不被诈骗商侵害？这些问题可通过各种安全协议解决。

➢ 电子现金。信用卡虽然方便，但不具有匿名性，而电子现金可以解决匿名性的问题。

➢ 游戏。怎么和跟你不在同一个房间的人玩抛硬币或纸牌游戏？如发牌就是一个问题。基于密码学思想可以解决这种问题。

2.2 古典密码学

古典密码的加密算法都基于两种运算：代替和置换。代替是将明文中的每个元素（如位、字母、位组或字母组等）映射成另一个元素，置换则是将明文中的元素重新排列。古典密码运算的基本要求是不允许有信息丢失（所有的运算是可逆的），大多数密码体制都使用了多层代替和置换。

古典密码本质上是将明文字母替换成其他字母、数字或符号的方法，比如把明文看成 0 或 1 的序列，那么密文就是 0 或 1 序列的另一种表达。

2.2.1 凯撒密码

凯撒密码是目前所知道的最早的代替密码，由 Julius Caesar 发明且运用于军事通信中。它的思想是将明文中的所有字母都在字母表上从前方（或后方）依据一个固定量偏移后作为密文。一般来说，明文使用小写字母，密文使用大写字母。例如，当偏移量为 3 时，对每个明文字母用其后的第 3 个字母代替，如表 2-1 所示。

表 2-1 凯撒密码加密映射表（密钥为 3）

明文	a	b	c	d	e	f	g	h	i	j	k	l	m	n	o	p	q	r	s	t	u	v	w	x	y	z
密文	D	E	F	G	H	I	J	K	L	M	N	O	P	Q	R	S	T	U	V	W	X	Y	Z	A	B	C

凯撒密码加密和解密可分为公式计算与查表两种方式。

➢ 公式加密

明文编码：令 $a=0, b=1, \cdots, z=25$，则

明文：$P = p_1 p_2 \cdots p_n$

（加密）运算：$C_i = p_i + k \pmod{26}$，$i = 1, 2, \cdots, n$

解码得密文：$C = c_1 c_2 \cdots c_n$

➢ 公式解密

密文：$C = c_1 c_2 \cdots c_n$

（解密）运算：$P_i = c_i - k \pmod{26}$，$i = 1, 2, \cdots, n$

解码得明文：$P = p_1 p_2 \cdots p_n$

➢ 查表加密（$K=5$）

当加密密钥为 5 时，可通过查表 2-2 实现加密。例如，假设明文为"iamfine"，则密文为"NFRKNSJ"。

表 2-2 凯撒密码加密映射表（密钥为 5）

明文	a	b	c	d	e	f	g	h	i	j	k	l	m	n	o	p	q	r	s	t	u	v	w	x	y	z
密文	F	G	H	I	J	K	L	M	N	O	P	Q	R	S	T	U	V	W	X	Y	Z	A	B	C	D	E

➢ 查表解密（$K=5$）

当加密密钥为 5 时，可通过查表 2-3 实现解密。例如，假设密文为"NFRKNSJ"，则明文为"iamfine"。

表 2-3 凯撒密码解密映射表（密钥为 5）

密文	A	B	C	D	E	F	G	H	I	J	K	L	M	N	O	P	Q	R	S	T	U	V	W	X	Y	Z
明文	v	w	x	y	z	a	b	c	d	e	f	g	h	i	j	k	l	m	n	o	p	q	r	s	t	u

➢ 凯撒密码分析

凯撒密码共有密钥 25 个，破解方可简单地依次去测试，利用强力搜索、穷举攻击的方式可破解密码，并且所破译的密码需要识别。不难看出，这样的密码加密形式的安全性较低，容易被破译。由于偏移量是固定的，所以最多只需尝试 25 种可能就能破译出凯撒密码。

2.2.2 希尔密码

希尔密码是运用基本矩阵原理的置换密码，由希尔（Lester S. Hill，1891—1961）发明于 1929 年。它是一种多表代替密码，矩阵原理上利用模运算意义下的矩阵乘法、求逆矩阵、线性无关、线性空间与线性变换等概念和运算。希尔密码有两个特点：一是字母的统计规律进一步降低，二是明文、密文字母不是一一对应关系。

希尔密码的基本思想是：将 d 个明文字母通过线性变换转换为 d 个密文字母。解密只要做一次逆变换就可以了，密钥就是变换矩阵本身。加密过程为：先对明文分组并编码，然后按公式 $C \equiv KP \bmod 26$ 计算，最后对计算结果 C 进行编码得到密文，其中，K 为密钥矩阵（要求存在逆矩阵），P、C 分别为明文、密文分组。解密过程为：先对密文分组并编码，然后利用公式 $P \equiv K^{-1}C \bmod 26$ 计算，最后对计算结果 P 进行编码得到明文，其中，K^{-1} 为密钥矩阵的逆矩阵。例如，若已知密钥矩阵为 $K = \begin{bmatrix} 8 & 21 \\ 21 & 1 \end{bmatrix}$，$K^{-1} = \begin{bmatrix} 3 & 15 \\ 15 & 24 \end{bmatrix}$，明文为"welovehainanuniversity"，则加解密过程如下。

由于密钥为 2×2 矩阵,故对明文按每 2 个字母为一组进行分组,$P_1=$"we",$P_2=$"lo",…,$P_{11}=$"ty"。下面先对 $P_1=$"we"进行加密:

$$P_1 = \begin{bmatrix} w \\ e \end{bmatrix} = \begin{bmatrix} 22 \\ 4 \end{bmatrix}$$

$$C_1 \equiv KP_1 \equiv \begin{bmatrix} 8 & 21 \\ 21 & 1 \end{bmatrix} \begin{bmatrix} 22 \\ 4 \end{bmatrix} = \begin{bmatrix} 260 \\ 466 \end{bmatrix} \equiv \begin{bmatrix} 0 \\ 24 \end{bmatrix} \mod 26$$

$$C_1 = \begin{bmatrix} 0 \\ 24 \end{bmatrix} = \begin{bmatrix} A \\ Y \end{bmatrix},\ 即 C_1 = "AY"。$$

同理,可得到其他明文分组的密文 $C_2, C_3, \cdots, C_{11}$,最后将它们组合在一起得到密文 $C=$"AYSLSDERZZNNRRLHZXAWGH"。

与加密时类似,解密时也要每 2 个字母为一组进行分组。下面以 $C_1=$"AY"为例说明解密过程:

$$C_1 = \begin{bmatrix} A \\ Y \end{bmatrix} = \begin{bmatrix} 0 \\ 24 \end{bmatrix}$$

$$P_1 \equiv K^{-1}C_1 \equiv \begin{bmatrix} 3 & 15 \\ 15 & 24 \end{bmatrix} \begin{bmatrix} 0 \\ 24 \end{bmatrix} = \begin{bmatrix} 360 \\ 576 \end{bmatrix} \equiv \begin{bmatrix} 22 \\ 4 \end{bmatrix} \mod 26$$

$$P_1 = \begin{bmatrix} 22 \\ 4 \end{bmatrix} = \begin{bmatrix} w \\ e \end{bmatrix},\ 即 P_1 = "we"。$$

2.2.3 栅栏密码

栅栏密码的思想是以列(行)优先写出明文,以行(列)优先读出各字母作为密文,换句话说,就是把要加密的明文分成 N 组,然后把每组的第 1 个字母连起来,形成一段无规律的密文,且组成栅栏的字母一般不会太多。栅栏密码的本质是置换,即通过打乱明文字母的排列顺序得到密文。

例如,要对明文消息"meet me at the park"采用栅栏密码进行加密,则可先按下列顺序写出明文:

```
m   e   m   a   t   e   a   k
 ↘ ↗ ↘ ↗ ↘ ↗ ↘ ↗ ↘ ↗ ↘ ↗ ↘ ↗
  e   t   e   t   h   p   r
```

然后按照行的顺序写出字母,即可得到密文:MEMATEAKETETHPR。

2.2.4 古典密码安全性分析

单表代替密码不是简单有序的字母移位,而是可以任意打乱字母的顺序,每个明文字母映射到一个不同的随机密文字母,看似安全(可有效应对穷举攻击),但实则不然。因为字母在单表代替密码的明文和密文中出现的概率是一致的,而人类语言的字母使用频率是固定已知的,英文字母频率和汉语拼音字母频率分别见图 2-1 和图 2-2。因此,单表代替密码不能抵御基于频率统计的攻击方法,故单表代替密码是极其不安全的,攻击者只需获得足够数量的密文,就可成功破解。

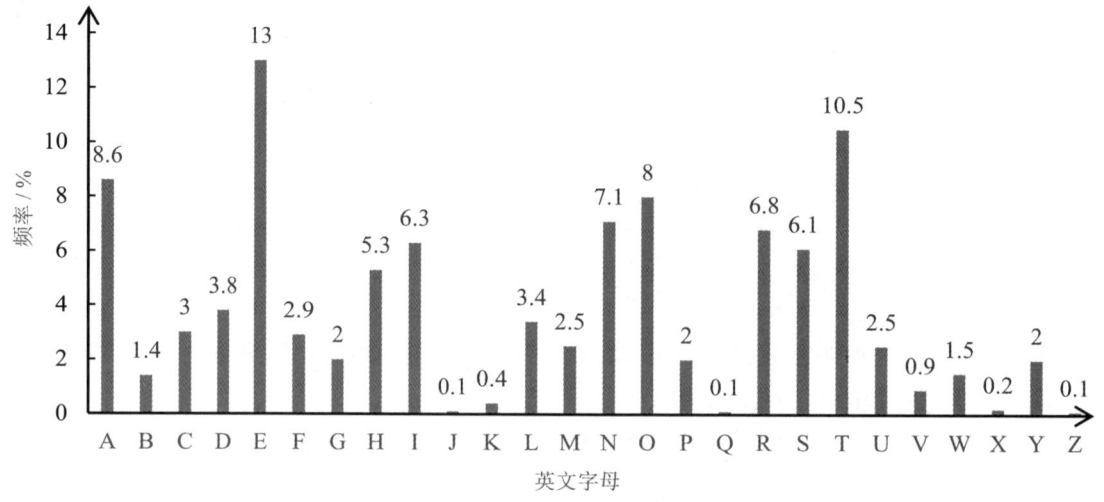

图 2-1 英文字母频率

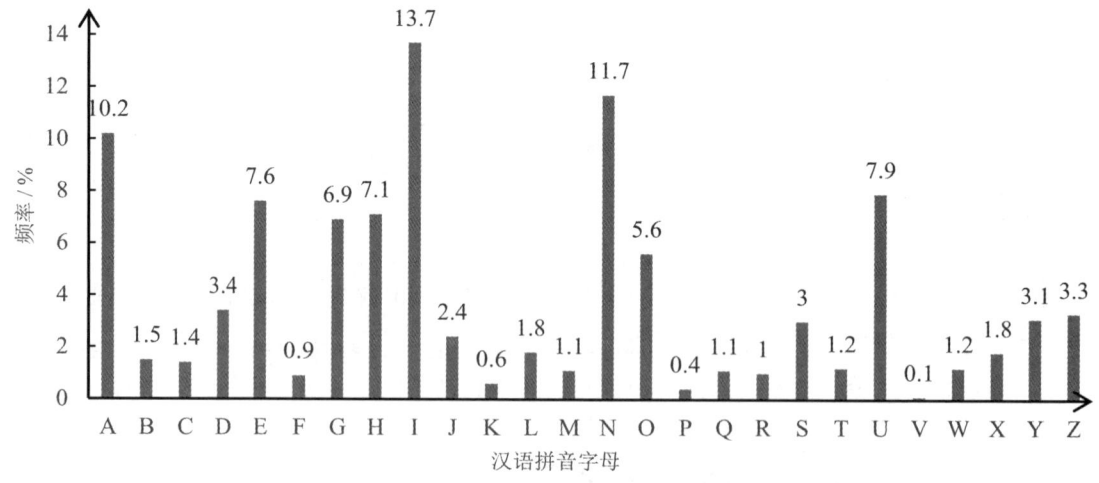

图 2-2 汉语拼音字母频率

多表代替密码在明文消息中采用不同的单表代换，可提高安全性，使得字母的频率分布更加均匀，用一个密钥指示明文消息中每个字母加/解密时所用的代替表，密钥和代替表可依次重复使用。虽然多表代替在一定程度上隐藏了明文字符的统计规律，但是随机性不强依然会被破解。

置换密码只是交换了明文字符的前后位置，而没有替换明文，因此保留了明文的较多信息，单纯使用置换对明文进行加密，其安全性并不高。

2.3 分组密码 DES

分组密码的一般设计原理是：将明文消息编码表示后的数字（简称明文数字）序列划分成长度为 n 的组（可看成长度为 n 的矢量），每组分别在密钥的控制下变换成等长的输出数字（简称密文数字）序列。

理想分组密码体制具有以下特点：分组密码作用在 n 位明文分组上，产生 n 位密文；共有 2^n 个不同的明文分组，可逆映射共 $2^n!$ 个；映射本身就是密钥，密钥长度为 $n \times 2^n$ 位，密钥规模大，难以直接应用。

针对理想分组密码存在的不足，Feistel 于 1967 年提出了所谓的 Feistel 网络加密模型，该模型通过交替地使用代替和置换技术，并基于乘积密码（依次使用两个或更多的基本密码）来逼近理想密码。Feistel 网络的加密密钥长为 k 位，分组长为 n 位，采用 2^k 个变换，而不是 $2^n!$ 个，方便实际应用。

和其他现代密码技术一样，分组密码 DES 为了挫败基于统计方法的密码攻击，采用了香农提出的基于混淆（confusion）和扩散（diffusion）的加密算法设计思想。其中，混淆的作用是使得密文的统计特性与密钥的取值之间的关系尽可能复杂；而扩散的作用是让明文的统计特征消散在密文中，使得明文和密文之间的统计关系尽可能复杂。

2.3.1 DES 轮迭代

美国国家标准局于 1973 年 5 月到 1974 年 8 月两次发布通告，公开征求用于电子计算机的加密算法。经评选，从一大批算法中采纳了 IBM 公司的 W. Tuchman 和 C. Meyer 提出的 LUCIFER 方案，该方案经标准化后于 1975 年 3 月公开发表，1977 年 1 月 15 日由美国国家标准局颁布为数据加密标准（Data Encryption Standard，DES），该标准于 1977 年 7 月 15 日生效。

DES 算法主要包括 2 次置换（初始置换和逆初始置换）、密钥控制下的 16 轮迭代加密和为每轮生成不同密钥的轮密钥发生器。

DES 对于每个 64 位的明文分组需要经历总共 16 轮的迭代加密，第 i 轮加密过程如图 2-3 所示。由该图可见，第 i-1 轮输出的右半部 32 位 R_{i-1} 及第 i 轮密钥 K_i 作为轮函数的输入，接着轮函数的输出 $f(R_{i-1}, K_i)$ 与第 i-1 轮输出的左半部 32 位 L_{i-1} 进行异或运算，异或运算的结果作为本轮迭代的右半部 32 位输出 R_i。而第 i-1 轮输出的右半部 32 位 R_{i-1} 未经处理直接作为本轮迭代的左半部 32 位输出 L_i。

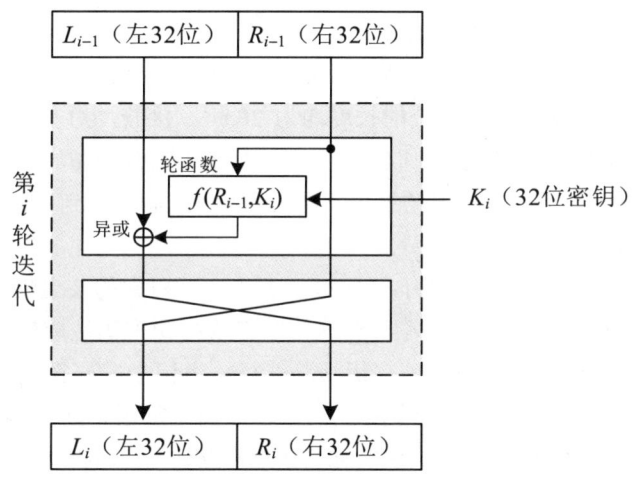

图 2-3　DES 第 i 轮加密过程

DES 轮函数的内部结构如图 2-4 所示，其主要处理过程包括：首先将上一轮右半部 32 位的 R_{i-1} 送入扩展 E 盒得到 48 位的结果，接着对该结果与 48 位的本轮密钥进行异或运算，然后对异或运算的结果进行 S 盒运算处理。S 盒运算在 DES 算法中是唯一的非线性运算，其他的运算全部是线性的。因此，S 盒在 DES 算法中提供了密码算法所必需的混淆作用，S 盒不易于分析，具有较高的安全性。但 S 盒的设计原理并未公开，不能从理论上证明其安全性，也因此受到密码界的质疑。S 盒一共包含 8 个不同的子盒，分别记为 $S_1, S_2, \cdots, S_8$。每个 S 子盒有 6 位输入和 4 位输出，因此 8 个 S 子盒最终实现了 48 位到 32 位的映射。S 盒运算可通过简单查表得到运算结果，每个 S 子盒是一个 4 行 16 列的表格，查表方法大致如下：从 6 位输入中提取出最高位和最低位，这两位按原来的高低位顺序组合在一起作为行号，剩余的中间 4 位作为列号；基于行号和列号到对应的 S 子盒进行查表，即可得到运算结果。

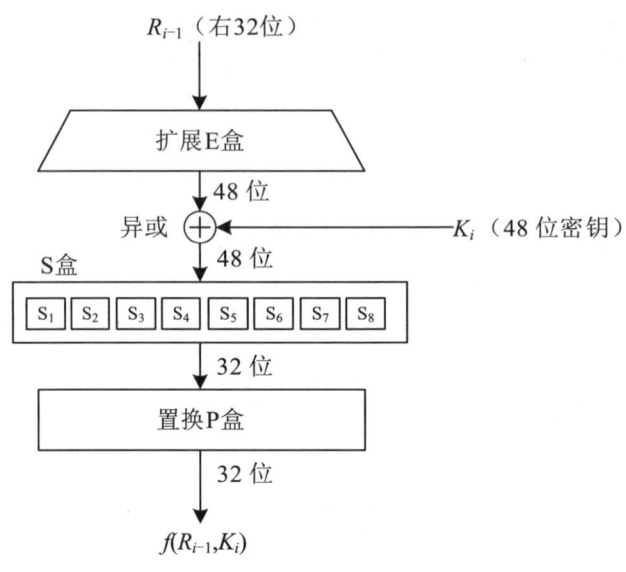

图 2-4　DES 轮函数的内部结构

2.3.2　DES 密钥生成及安全性

　　DES 的轮密钥生成器的主要操作过程如下：由用户输入的 64 位种子密钥经过压缩置换 P1 盒处理（去掉每个字节的第 8 位奇偶校验位，实现 64 位密钥压缩置换至 56 位）得到 56 位的密钥，将此密钥等分成左右两部分，每部分的长度均为 28 位。接着，对左右两部分单独进行循环左移 n 位（当轮次为 1、2、9、16 时，n 取 1，否则 n 取 2），合并循环左移 n 位后的左右两部分得到 56 位的结果。最后，将 56 位的结果送入压缩置换 P2 盒，得到 48 位的第 1 轮加密密钥。类似地，可得到其他各轮的子密钥。

　　DES 的安全性完全依赖于所用的密钥。从 DES 诞生起，对它的安全性就有激烈的争论，一直延续到现在。由前面可知，DES 算法在每轮迭代时都要由轮密钥生成器提供一个子密钥，以便在轮迭代加密时使用。对于给定的种子密钥 k，若经轮密钥生成器得到的各轮子密钥都相同，即有 $k1=k2=\cdots=k16$，则称给定的种子密钥 k 为弱密钥。使用密钥 k 对 x 加密或解密两次，均可恢复出明文。因此，当 k 为弱密钥时，DES 的加密运算和解密运算是一样的，没有任何区别。

　　一般来说，弱密钥的存在并不会从根本上危及 DES 算法的安全性。但由于 DES 的密钥空间太小，按 1999 年时的算力，只需 22 小时就可完成对 DES 密码的穷举攻击，因此在 20 世纪 90 年代，DES 已经退出了商用密码的舞台。

2.4　分组密码 AES

　　高级加密标准（Advanced Encryption Standard，AES）是由比利时密码专家 Vincent Rijmen

和 Joan Daemen 设计出来的对称加密算法，在密码学中又称为 Rijndael 加密法，是美国联邦政府采用的一种分组加密标准。AES 的设计目标是力求满足以下三个标准：①能够有效抵抗所有已知的攻击；②能够运行于各种不同平台上且加/解密速度快；③设计简单，编码紧凑，易于实现。

AES 算法采用的不是 Fesitel 网络结构，在每轮迭代时可以基于线性层和非线性层并行处理整个明文或密文分组，其扩散速度比采用 Fesitel 网络结构的 DES 算法更快。为了更好地应对穷举攻击，AES 可以选用三种不同长度的密钥，分别是 128 位、192 位和 256 位。

2.4.1 AES 轮迭代

AES 的基本处理块的大小为 1 字节，128 位的明文分组输入 AES 加密端后，会被看成 16 字节，并按照先列后行的顺序存储在一个 4 行 4 列的数组中，此数组称为状态数组。类似地，128 位的种子密钥也存储在一个 4 行 4 列的数组中，密钥生成器基于种子密钥生成 $N+1$ 个子密钥供初始轮密钥加和 N 轮迭代中的轮密钥加使用。迭代轮次 N 的值与密钥的长度有关，AES 规定：当密钥长度分别为 128 位、192 位、256 位时，N 分别取为 10、12 和 14，本节均取 $N=10$。

AES 的每轮迭代（最后一轮除外）包括四部分：非线性的字节代替（Byte Sub，BS）、线性的行移位（Shift Row，SR）、线性的列混淆（Mix Column，MC）和线性的轮密钥加（Add Round Key，ARK）。最后一轮只包括字节代替、行移位和轮密钥加三部分。

1. 字节代替

字节代替是 AES 算法中唯一的非线性变换，对状态数组中的每个元素依次进行代替，加密时可通过简单查表（S 盒，见表 2-4）实现字节代替；解密时需要查 S 逆盒（表 2-5）。具体方法是：元素用十六进制数表示，其高位值当作行号（x），其低位值当作列号（y），按行列号查 AES 的 S 盒或 S 逆盒，得到字节代替的值。例如，状态数组中的某个元素为 6A，若要加密，则查 S 盒的 6 行 A 列，得到 02；若要解密，则查 S 逆盒的 6 行 A 列，得到 58。

表 2-4 AES 的 S 盒

x	\multicolumn{16}{c}{y}															
	0	1	2	3	4	5	6	7	8	9	A	B	C	D	E	F
0	63	7C	77	7B	F2	6B	6F	C5	30	01	67	2B	FE	D7	AB	76
1	CA	82	C9	7D	FA	59	47	F0	AD	D4	A2	AF	9C	A4	72	C0
2	B7	FD	93	26	36	3F	F7	CC	34	A5	E5	F1	71	D8	31	15
3	04	C7	23	C3	18	96	05	9A	07	12	80	E2	EB	27	B2	75
4	09	83	2C	1A	1B	6E	5A	A0	52	3B	D6	B3	29	E3	2F	84
5	53	D1	00	ED	20	FC	B1	5B	6A	CB	BE	39	4A	4C	58	CF
6	D0	EF	AA	FB	43	4D	33	85	45	F9	02	7F	50	3C	9F	A8
7	51	A3	40	8F	92	9D	38	F5	BC	B6	DA	21	10	FF	F3	D2
8	CD	0C	13	EC	5F	97	44	17	C4	A7	7E	3D	64	5D	19	73
9	60	81	4F	DC	22	2A	90	88	46	EE	B8	14	DE	5E	0B	DB
A	E0	32	3A	0A	49	06	24	5C	C2	D3	AC	62	91	95	E4	79
B	E7	C8	37	6D	8D	D5	4E	A9	6C	56	F4	EA	65	7A	AE	08
C	BA	78	25	2E	1C	A6	B4	C6	E8	DD	74	1F	4B	BD	8B	8A

续表

x	\multicolumn{16}{c}{y}
	0	1	2	3	4	5	6	7	8	9	A	B	C	D	E	F
D	70	3E	B5	66	48	03	F6	0E	61	35	57	B9	86	C1	1D	9E
E	E1	F8	98	11	69	D9	8E	94	9B	1E	87	E9	CE	55	28	DF
F	8C	A1	89	0D	BF	E6	42	68	41	99	2D	0F	B0	54	BB	16

表 2-5 AES 的 S 逆盒

x	\multicolumn{16}{c}{y}
	0	1	2	3	4	5	6	7	8	9	A	B	C	D	E	F
0	52	09	6A	D5	30	36	A5	38	BF	40	A3	9E	81	F3	D7	FB
1	7C	E3	39	82	9B	2F	FF	87	34	8E	43	44	C4	DE	E9	CB
2	54	7B	94	32	A6	C2	23	3D	EE	4C	95	0B	42	FA	C3	4E
3	08	2E	A1	66	28	D9	24	B2	76	5B	A2	49	6D	8B	D1	25
4	72	F8	F6	64	86	68	98	16	D4	A4	5C	CC	5D	65	B6	92
5	6C	70	48	50	FD	ED	B9	DA	5E	15	46	57	A7	8D	9D	84
6	90	D8	AB	00	8C	BC	D3	0A	F7	E4	58	05	B8	B3	45	06
7	D0	2C	1E	8F	CA	3F	0F	02	C1	AF	BD	03	01	13	8A	6B
8	3A	91	11	41	4F	67	DC	EA	97	F2	CF	CE	F0	B4	E6	73
9	96	AC	74	22	E7	AD	35	85	E2	F9	37	E8	1C	75	DF	6E
A	47	F1	1A	71	1D	29	C5	89	6F	B7	62	0E	AA	18	BE	1B
B	FC	56	3E	4B	C6	D2	79	20	9A	DB	C0	FE	78	CD	5A	F4
C	1F	DD	A8	33	88	07	C7	31	B1	12	10	59	27	80	EC	5F
D	60	51	7F	A9	19	B5	4A	0D	2D	E5	7A	9F	93	C9	9C	EF
E	A0	E0	3B	4D	AE	2A	F5	B0	C8	EB	BB	3C	83	53	99	61
F	17	2B	04	7E	BA	77	D6	26	E1	69	14	63	55	21	0C	7D

下面简要说明 S 盒的构造过程。

首先,构造一个 16 行 16 列的二维表格,行列号用十六制数从 0 至 F 依次编号。表格各单元格的初始值就是该单元所对应的行列号,即在 x 行 y 列的单元格的字节值是"xy"。完成初始化后的表格的各行字节值为:第一行是 00, 01, 02, …, 0F;第二行是 10, 11, 12, …, 1F;……最后一行是 F0, F1, F2, …, FF。

其次,将初始化后的每字节映射成其在有限域 $GF(2^8)$ 中的逆,特别地,约定"00"映射为它自身——00。

最后,将 S 盒中映射后的每字节的 8 位记为 $(b_7 b_6 b_5 b_4 b_3 b_2 b_1 b_0)$,则变换后的字节 8 位值 $(b'_7 b'_6 b'_5 b'_4 b'_3 b'_2 b'_1 b'_0)$ 由下面的矩阵运算得到:

$$\begin{bmatrix} b'_7 \\ b'_6 \\ b'_5 \\ b'_4 \\ b'_3 \\ b'_2 \\ b'_1 \\ b'_0 \end{bmatrix} = \begin{bmatrix} 1 & 1 & 1 & 1 & 1 & 0 & 0 & 0 \\ 0 & 1 & 1 & 1 & 1 & 1 & 0 & 0 \\ 0 & 0 & 1 & 1 & 1 & 1 & 1 & 0 \\ 0 & 0 & 0 & 1 & 1 & 1 & 1 & 1 \\ 1 & 0 & 0 & 0 & 1 & 1 & 1 & 1 \\ 1 & 1 & 0 & 0 & 0 & 1 & 1 & 1 \\ 1 & 1 & 1 & 0 & 0 & 0 & 1 & 1 \\ 1 & 1 & 1 & 1 & 0 & 0 & 0 & 1 \end{bmatrix} \begin{bmatrix} b_7 \\ b_6 \\ b_5 \\ b_4 \\ b_3 \\ b_2 \\ b_1 \\ b_0 \end{bmatrix} \oplus \begin{bmatrix} 0 \\ 1 \\ 1 \\ 0 \\ 0 \\ 0 \\ 1 \\ 1 \end{bmatrix} \quad (2\text{-}1)$$

2. 行移位

每轮迭代的行移位是线性的。AES 加密时，状态数组的第一行不移位，第二行循环左移 1 位，第三行循环左移 2 位，第四行循环左移 3 位，如图 2-5 所示。显然，行移位是可逆的，即当 AES 解密时，状态数组的第一行不移位，第二行循环右移 1 位，第三行循环右移 2 位，第四行循环右移 3 位，如图 2-6 所示。

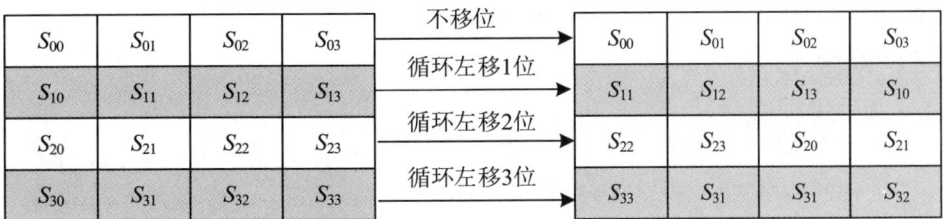

图 2-5 AES 行移位变换

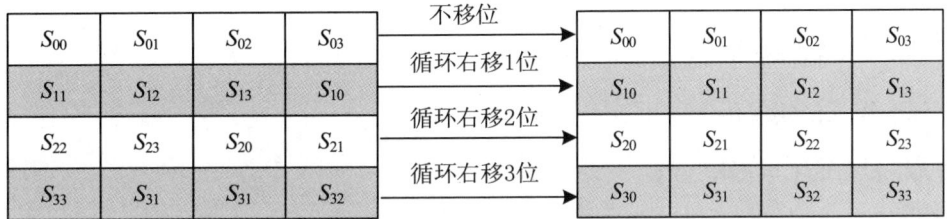

图 2-6 AES 逆行移位变换

3. 列混淆

列混淆属于线性变换，在轮迭代中对状态数组的每列分别进行代替操作。状态数组中的每字节被映射为一个新的值，该值由同一列中的 4 字节通过下面的数组运算得到：

$$\begin{bmatrix} S'_{00} & S'_{01} & S'_{02} & S'_{03} \\ S'_{10} & S'_{11} & S'_{12} & S'_{13} \\ S'_{20} & S'_{21} & S'_{22} & S'_{23} \\ S'_{30} & S'_{31} & S'_{32} & S'_{33} \end{bmatrix} = \begin{bmatrix} 02 & 03 & 01 & 01 \\ 01 & 02 & 03 & 01 \\ 01 & 01 & 02 & 03 \\ 03 & 01 & 01 & 02 \end{bmatrix} \begin{bmatrix} S_{00} & S_{01} & S_{02} & S_{03} \\ S_{10} & S_{11} & S_{12} & S_{13} \\ S_{20} & S_{21} & S_{22} & S_{23} \\ S_{30} & S_{31} & S_{32} & S_{33} \end{bmatrix} \tag{2-2}$$

逆列混淆用于解密迭代轮次中。类似地，逆列混淆也要将状态数组中的每字节映射为一个新的值，该值由同一列中的 4 字节通过下面的数组运算得到：

$$\begin{bmatrix} S'_{00} & S'_{01} & S'_{02} & S'_{03} \\ S'_{10} & S'_{11} & S'_{12} & S'_{13} \\ S'_{20} & S'_{21} & S'_{22} & S'_{23} \\ S'_{30} & S'_{31} & S'_{32} & S'_{33} \end{bmatrix} = \begin{bmatrix} 0E & 0B & 0D & 09 \\ 09 & 0E & 0B & 0D \\ 0D & 09 & 0E & 0B \\ 0B & 0D & 09 & 0E \end{bmatrix} \begin{bmatrix} S_{00} & S_{01} & S_{02} & S_{03} \\ S_{10} & S_{11} & S_{12} & S_{13} \\ S_{20} & S_{21} & S_{22} & S_{23} \\ S_{30} & S_{31} & S_{32} & S_{33} \end{bmatrix} \tag{2-3}$$

4. 轮密钥加

AES 的轮密钥加就是将轮子密钥与状态数组简单地按位进行异或运算。

2.4.2 轮密钥生成器

AES 的轮密钥生成器根据种子密钥生成加密和解密所需的 $N+1$ 个子密钥，由于本书只针对子密钥长度为 128 位的情况,密钥生成器总共需要生成 $11\times128b=11\times4\times32b=44\times4B$,因此可用一个一维数组 W[44] 来存储扩展密钥,数组元素的长度单位为字(一个数组元素的长度为 4 字节)。轮密钥生成器的密钥扩展算法的实现代码如下:

```
KeyExpansion (byte Key[16] , W[44]) {
  for (i =0; i < 4; i ++)   // W[0]至 W[3]由种子密钥生成
    W[i]=(Key[4*i],Key[4*i +1],Key[4*i +2],Key[4*i +3] );
      for (i =4; i <44; i ++) {
        g=W[i-1];
        if (i % 4= =0) //在 i 整除 4 的位置,给 g 赋新值
          g=SubByte (RotByte (g))^Rcon[i /4];
        W[i]=W[i-4]^ g;}
}
```

2.4.3　AES 算法实现

AES 加密算法的实现伪代码如下:

```
Encryption(State,CipherKey) {
  KeyExpansion(CipherKey, RoundKey);
  AddRoundKey(State, RoundKey);
  for(i=1;i<Nr;i++)
    Round(State, RoundKey) {
      ByteSub(State);
      ShiftRow(State);
      MixColumn(State);
      AddRoundKey(State, RoundKey);}
    FinalRound(State, RoundKey) {
      ByteSub(State);
      ShiftRow(State);
      AddRoundKey(State, RoundKey);}
}
```

AES 解密算法的实现伪代码如下:

```
Decryption(State,CipherKey) {
  Inv_KeyExpansion(CipherKey,Inv_ RoundKey);
  AddRoundKey(State,Inv_ RoundKey);
  for(i=1;i<Nr;i++)
    Inv_Round(State, Inv_ RoundKey) {
      Inv_ShiftRow(State);
      Inv_ByteSub(State);
```

```
        AddRoundKey(State, Inv_ RoundKey;
            Inv_MixColumn(State);}
    Inv_FinalRound(State,Inv_RoundKey) {
        InvShiftRow(State);
        InvByteSub(State);
        AddRoundKey(State, Inv_ RoundKey );}
}
```

2.5 非对称密码体制

目前常用的非对称密码主要包括 RSA 和椭圆曲线密码。

RSA 算法由 Ron Rivest、Adi Shamir 和 Len Adleman 于 1977 年在麻省理工学院开发完成，是目前应用最广泛的一种商用公钥加密方法，其明文和密文是介于 0 和 n 之间的整数，n 的典型值是 1024 位二进制数（约 309 个十进制数）。

RSA 使用具有指数的表达式，明文以分组的形式加密，每个分组的二进制值小于某个指定数字 n。加/解密的形式如下：对于明文块 M 和密文块 C，发送方和接收方都必须知道 n 的值，发送方知道 e 的值，只有接收方知道 d 的值，这是一种公钥加密算法，公钥 PU=$\{e,n\}$，私钥 PR=$\{d,n\}$。

要使该算法满足公钥加密的要求，必须满足以下要求：

（1）易于找到 e、d、n 的值，使得 M^{ed} mod $n = M$ 对于所有 $M < n$ 成立；

（2）对于 $M < n$ 的所有值，计算 M^e mod n 和 C^d mod n 相对容易；

（3）给定 e 和 n，确定 d 在计算上是不可行的。

椭圆曲线加密算法（ECC）是基于椭圆曲线数学理论实现的一种非对称加密算法。相比 RSA，ECC 的优势是可以使用更短的密钥，实现与 RSA 相当或更高的安全性。

第 3 章 防火墙技术

3.1 防火墙概述

顾名思义，广义上的防火墙（firewall）就是隔断火患和财物之间的一堵墙，以此来达到减少财物损失的目的。而计算机领域中的防火墙的功能就像现实中的防火墙一样，把绝大多数的外来侵害挡在外面，保护内部计算机的安全。

3.1.1 防火墙的基本概念

防火墙通常是指设置在不同网络（如可信任的企业内部网和不可信任的公共网）或网络安全域之间的一系列部件的组合（包括硬件和软件）。它是不同网络或网络安全域之间信息的唯一出入口，能根据企业的安全政策控制（允许、拒绝、监测）出入网络的信息流，且本身具有较强的抗攻击能力。它是提供信息安全服务、实现网络和信息安全的基础设施。

在逻辑上，防火墙是一个分离器，是一个限制器，也是一个分析器，可有效地监控内部网络和外部网络（如互联网）之间的通信流量，保证内部网络的安全。防火墙的逻辑位置如图 3-1 所示。

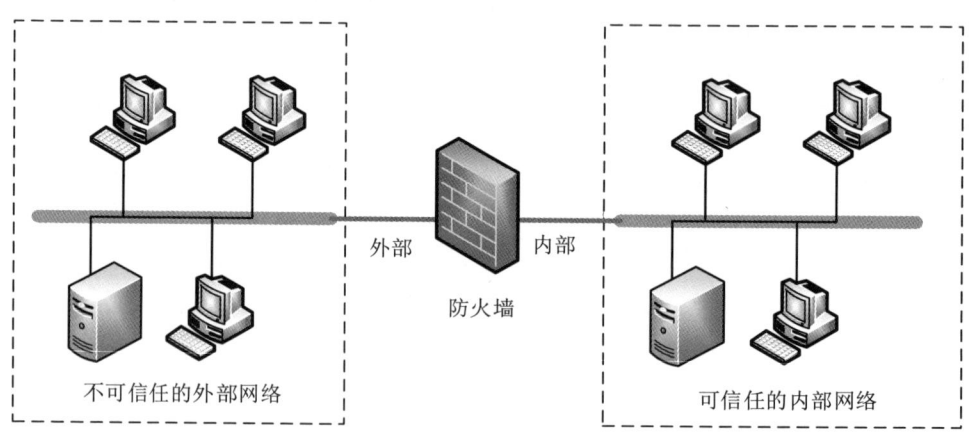

图 3-1 防火墙的逻辑位置

由于防火墙设定了网络边界和服务，因此更适用于相对独立的网络，例如内联网（Intranet）等。防火墙是非常流行的控制访问网络系统的方法。事实上，在互联网上，超过三分之一的Web 网站都是由某种形式的防火墙加以保护的，这是对黑客防范最严格、安全性较强的一种方式。任何关键性的服务器都应放在防火墙之后。

3.1.2 防火墙的功能

防火墙能增强内部网络的安全性，加强网络间的访问控制，防止外部用户非法使用内部网络资源，保护内部网络不被破坏，防止内部网络的敏感数据被窃取。防火墙可决定外界可以访问哪些内部服务，以及内部人员可以访问哪些外部服务。一般来说，防火墙应该具备以下功能：

- 支持安全策略。即使在没有其他安全策略的情况下，也应该支持"除非特别许可，否则拒绝所有的服务"的设计原则。
- 易于扩充新的服务和更改所需的安全策略。
- 具有代理服务功能（如 FTP、Telnet 等），包含先进的鉴别技术。
- 采用过滤技术，根据需求允许或拒绝某些服务。
- 具有灵活的编程语言，界面友好，且具有很多过滤属性，包括源和目的 IP 地址、协议类型、源和目的 TCP/UDP 端口。
- 具有缓冲存储的功能，以提高访问速度。
- 能够接纳对内部网络的公共访问，对内部网络的公共信息服务进行保护，并根据需要删减或扩充。
- 具有对拨号访问内部网络的集中处理和过滤能力。
- 具有记录和审计功能，便于检查和审计。
- 防火墙设备上所使用的操作系统和开发工具都应该具备相当等级的安全性。
- 可检验和可管理。

3.2 防火墙的分类

3.2.1 分组过滤型防火墙

分组过滤也称为包过滤，是一种通用、廉价、有效的安全手段。

包过滤在网络层和传输层起作用。它根据分组（包）的源、目的地址，端口号及协议类型，标志（来源于 IP、TCP 或 UDP 包头）确定是否允许分组（包）通过。

包过滤的优点是不用改动主机上的应用程序，因为它工作在网络层和传输层，与应用层无关。但其弱点也是明显的：由于过滤判别的只有网络层和传输层的有限信息，因此不可能充分满足各种安全要求；在许多过滤器中，过滤规则的数目是有限制的，且随着规则数目的增大，性能会受到很大的影响；由于缺少上下文关联信息，不能有效地过滤如 UDP、RPC 一类的协议；另外，大多数过滤器中缺少审计和报警机制，且管理方式和用户界面较差；对安全管理人员素质要求高，建立安全规则时，必须对协议本身及其在不同应用程序中的作用有较深入的理解。因此，过滤器通常和应用网关配合使用，共同组成防火墙系统。

3.2.2 应用代理型防火墙

应用代理型防火墙是内部网络与外部网络的隔离点，起着监视和隔绝应用层通信流的作用。它工作在 OSI 模型的最高层，管控着应用系统中可用于安全决策的全部信息。

应用代理型防火墙的优点：不允许数据包直接通过防火墙，避免了数据驱动式攻击的发生，安全性好；能生成各项记录；能灵活、完整地控制进出的流量和内容；能过滤数据内容。应代理型防火墙的缺点则表现在：对于每项服务代理可能要求不同的服务器；速度较慢；对用户不透明，用户需要改变客户端程序；不能保证免受所有协议弱点的限制；不能提升底层协议的安全性。

3.2.3 复合型防火墙

由于对更高安全性的要求，常把基于包过滤的方法与基于应用代理的方法结合起来，形成复合型防火墙产品。这种结合通常有以下两种结构。

屏蔽主机防火墙体系结构：在该结构中，分组过滤路由器或防火墙与互联网相连，同时一个堡垒主机安装在内部网络，通过在分组过滤路由器或防火墙上过滤规则的设置，使堡垒主机成为互联网上其他节点所能到达的唯一节点，这确保了内部网络不受未授权外部用户的攻击。

屏蔽子网防火墙体系结构：堡垒主机放在一个子网内，形成非军事化区，两个分组过滤路由器放在这一子网的两端，使这一子网与互联网及内部网络分离。在屏蔽子网防火墙体系结构中，堡垒主机和分组过滤路由器共同构成了整个防火墙的安全基础。

3.3 防火墙的结构

防火墙既可以是一台路由器、一台个人计算机（PC）或者一台主机，也可以是由多台主机构成的体系。防火墙结构包括堡垒主机、主机驻留防火墙、网络设备防火墙、个人防火墙等。

3.3.1 堡垒主机

"堡垒"一词起源于中世纪，是指城堡中特别加固的部分，用于发现和抵御攻击者的进攻。在网络中，堡垒主机是安装了防火墙软件，但没有 IP 转发功能的计算机。它对外界提供一些必要的服务，也可以被内部用户访问。通常它只提供一种服务，因为提供的服务越多，导致安全隐患的可能性也就越大。它应该位于 DMZ（DeMilitarized Zone，非军事区，也称为停火区或者周边网络）中。如果堡垒主机提供代理服务，它会知道自己将要为哪些应用提供代理。常见的堡垒主机可以分为以下 5 种类型。

单宿主堡垒主机：有一块网卡的堡垒主机做防火墙，通常用于应用级网关防火墙，将外部路由器配置成所有进来的数据均发送到堡垒主机上，同时将全部内部客户端配置成所有出去的数据都发送到这台堡垒主机上。堡垒主机以安全方针为依据检验这些数据。其主要缺点是可以配置路由器使信息直接进入内部网络，而完全绕过堡垒主机；内部用户也可以配置他们的主机，绕过堡垒主机把信息直接发送到路由器上。

双宿主堡垒主机：有两块网卡的堡垒主机做防火墙，两块网卡各自与内外部网络相连，但是内外部网络之间不能直接通信，内外部网络之间的数据流被双宿主机完全切断。它采用主机取代路由器执行安全控制功能，可以通过运行代理软件或者让用户直接注册到其上来提供网络控制。当黑客访问内部网络时，必须首先攻破双宿主堡垒主机，这使得网络管理员有时间对入侵做出反应。

内部堡垒主机：堡垒主机与内部网络通信，以便转发从外部网络获得的信息。这类堡垒主机启用了较多的服务，并开放了较多的端口，以满足应用程序的需要。

外部堡垒主机：堡垒主机为互联网提供公共服务，它不向内部网络转发任何请求，而是自己处理请求。它只提供非常有限的服务，并且只开放有限的端口来满足这类服务。它需要更多的防御和保护，并应切断对内部网的任何访问。

受害堡垒主机：该类堡垒主机是故意向攻击者暴露的目标，也被称作蜜罐或者陷阱。设置它的主要目的是引诱黑客的攻击，让黑客误以为已经成功入侵网络，并且让黑客继续"为所欲为"，以便赢得时间跟踪他们。该类堡垒主机只包含最起码的最少服务配置，以便运行相应的程序。

3.3.2 主机驻留防火墙

主机驻留防火墙是一个保障个人主机安全的软件模块，一般是系统自带或作为附件出现的，通常位于服务器上，主要用来保护主机，以免其受到网络攻击。主机驻留防火墙的功能如下：控制主机的程序是否可以访问网络，未经过授权的程序不准访问网络；防止攻击者进行端口扫描和利用开放端口进行攻击。

主机驻留防火墙具有以下优点：过滤规则可以根据主机环境定制，保护功能独立于网络拓扑结构，为系统提供了一个额外的保护层。

3.3.3 网络设备防火墙

网络设备防火墙是一种利用网络设备对外部网络进行过滤的防火墙，通常在网络设备中监视和过滤通过设备的数据包流，最具代表性的例子为屏蔽路由器。

屏蔽路由器是在互联网和内部网络之间放置的路由器，执行包过滤功能，这是最简单的一种防火墙。屏蔽路由器是内外连接的唯一通道，所有的数据包都必须在此通过检查。在路由器上安装包过滤软件，实现包过滤功能。图 3-2 显示了它的拓扑结构，虽然它并不昂贵，但能提供重要的保护。屏蔽路由器体系结构也称筛选路由器体系结构，最大优点是架构简单且硬件成本较低，由于路由器提供非常有限的服务，因此保护路由器比保护主机更易实现。

图 3-2　具有包过滤功能的路由器

3.3.4 个人防火墙

个人防火墙是安装在个人计算机系统里的一段"代码墙"，把个人计算机和互联网隔开，能够控制个人计算机、工作站与互联网或企业网络之间的网络流量。它检查到达防火墙两端的所有数据包，无论是进入的还是发出的，决定拦截还是放行。也就是说，在不妨碍你正常上网浏览的同时，阻止互联网上的其他用户对你的计算机进行非法访问。

个人防火墙的主要功能是按照给定的安全规则处理网络数据包，并记录安全日志。一个好的个人防火墙必须系统资源消耗少，处理效率高，具有简单易懂的设置界面，具有灵活而有效的规则设定。

常见的个人防火墙有天网防火墙个人版、瑞星个人防火墙、360 木马防火墙、费尔个人防火墙、江民黑客防火墙和金山网标等。

3.4 防火墙的部署

防火墙的部署包括 DMZ 的部署、VPN 的部署和分布式防火墙的部署。

3.4.1 DMZ 的部署

DMZ 是为了解决安装防火墙后外部网络不能访问内部网络服务器的问题而设立的一个非安全系统与安全系统之间的缓冲区。DMZ 位于企业内部网络和外部网络之间，该区域内可以放置一些必须对外开放的服务器，如企业 Web 服务器、FTP 服务器等。通过 DMZ，可以更加有效地保护内部网络，因为这种部署比一般的防火墙方案，对来自外部网络的攻击多了一道防线。

通常按照功能或者部门将网络分割成若干网段，不同网段往往有着不同的安全需求。以太网是一个广播式的网络，在同一个网段内的主机都可以接收到该网络的所有广播分组。如果黑客入侵网络，可以容易地截获相应主机间传输的分组。为了配置和管理方便，内部网络中需要向外提供服务的服务器往往放在一个单独的网段，这个网段便是 DMZ。DMZ 在内部网络之外，具有一个与内部网络不同的网络号，连接到防火墙，提供公共服务。

可以采用多种不同的方法部署 DMZ，下面介绍三种典型的方法。

1. DMZ 位于单防火墙接口之下

使用一个有三个接口的防火墙（三宿主防火墙）创建隔离区，如图 3-3 所示，每个隔离区连接一个防火墙接口。防火墙提供区之间的隔离，也有助于 DMZ 的安全。

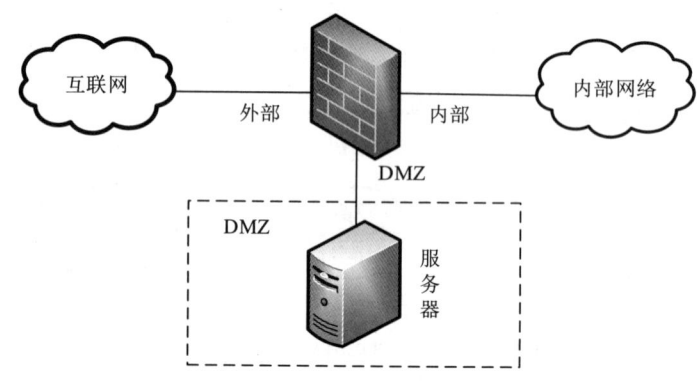

图 3-3　DMZ 位于单防火墙接口之下

2. DMZ 位于防火墙之外

如图 3-4 所示，DMZ 位于防火墙之外，不在互联网和防火墙之间的通道上。DMZ 位于边缘路由器的一个接口下，没有与防火墙直接相连，从 DMZ 到防火墙形成一个隔离层。在这种配置中，路由器能够拒绝所有从 DMZ 子网到防火墙所在的子网的访问，当位于 DMZ 子网的主机受到危害且攻击者开始使用这个主机对网络发动进一步攻击时，增加的隔离层能够延缓对防火墙的攻击进度。

3. DMZ 位于两个防火墙之间

如图 3-5 所示，DMZ 位于两个防火墙之间，由两个防火墙共同来保护系统的安全。外部防火墙监控 DMZ 与互联网之间的通信，内部防火墙监控 DMZ 与内部网络之间的通信。

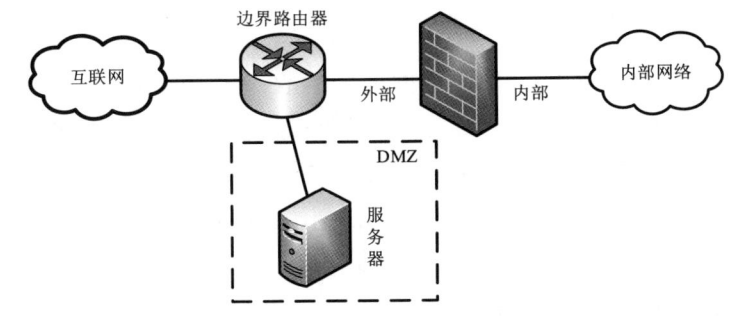

图 3-4　DMZ 位于防火墙之外

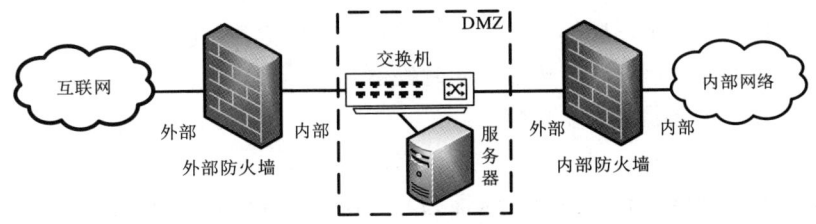

图 3-5　DMZ 位于两个防火墙之间

这种方法使 DMZ 获得了相当高的安全性，但缺点也有两个：一个是从互联网到内部网络的所有流量都必须通过 DMZ，从内部网络到互联网的所有流量也都要经过 DMZ，一个 DMZ 设备被攻陷后，攻击者会阻截这些流量，解决的办法是在两个防火墙之间的设备上使用虚拟局域网（VLAN）；另一个是需要使用两个防火墙，增加了设备成本。

大型公司或企业通常使用基于两个防火墙的体系结构来提高网络的安全性，某大型国有企业的防火墙配置拓扑如图 3-6 所示。

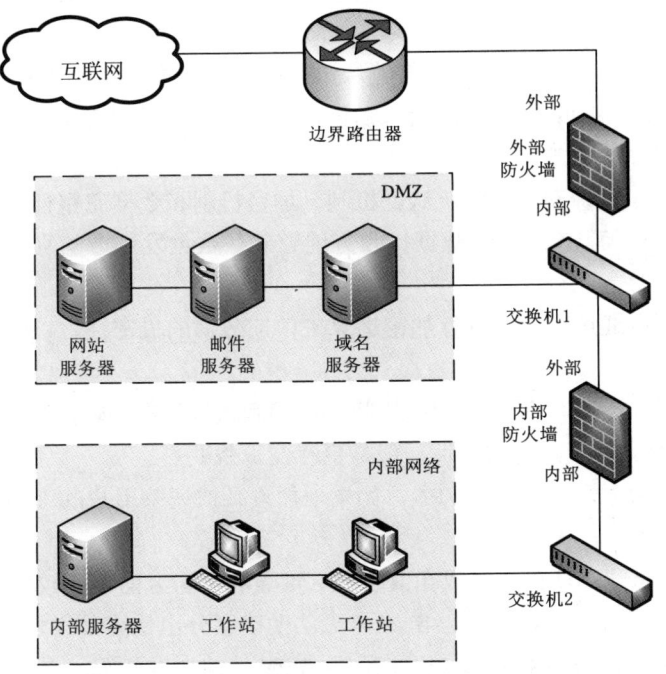

图 3-6　某大型国有企业的防火墙配置拓扑

3.4.2　VPN 的部署

VPN（Virtual Private Network，虚拟私有网络）是指在公用网络上建立的私有的、专用的虚拟通信网络。VPN 在企业网络中有广泛的应用，VPN 网关通过对数据包的加密和数据包目的地址的转换实现远程访问。VPN 有多种分类方式，按照 VPN 技术实现的网络层次，可以进行如下分类。

> 基于数据链路层的 VPN：L2TP VPN、L2F VPN、PPTP VPN。
> 基于网络层的 VPN：GRE VPN、IPsec VPN。
> 基于应用层的 VPN：SSL VPN。

下面以 IPsec VPN 为例说明 VPN 防火墙的部署与配置。IPsec（Internet Protocol security，互联网络层安全协议）是一种三层隧道协议。三层隧道协议把各种网络协议直接装入隧道协议中，形成的数据包依靠第三层协议进行传输。同时提供安全协议选择、安全算法，确定服务所使用密钥等服务，从而在 IP 层提供安全保障。

IPsec 是一个框架性架构，具体由两类协议组成。

（1）AH（Authentication Header，认证头）协议：可以同时提供数据完整性确认、数据来源确认、防重放等安全特性。AH 常用摘要算法（单向 Hash 函数）MD5 和 SHA1 实现这些安全特性。

（2）ESP（Encapsulated Security Payload，封装安全载荷）协议：可以同时提供数据完整性确认、数据加密、防重放等安全特性。ESP 通常使用 DES、3DES、AES 等加密算法实现数据加密，使用 MD5 或 SHA1 实现数据完整性。

IPsec 有如下两种封装模式。

（1）隧道（tunnel）模式：用户的整个 IP 数据包被用来计算 AH 或 ESP 头，AH 或 ESP 头以及 ESP 加密的用户数据被封装在一个新的 IP 数据包中。通常，隧道模式应用于两个安全网关之间的通信。

（2）传输（transport）模式：只有传输层数据被用来计算 AH 或 ESP 头，AH 或 ESP 头以及 ESP 加密的用户数据被放置在原 IP 首部后面。通常，传输模式应用于两台主机之间的通信或一台主机和一个安全网关之间的通信。

IPsec VPN 提供了三种安全机制：认证机制、加密机制和数据完整性。

（1）认证机制：使 IP 通信的数据接收方能够确认数据发送方的真实身份以及数据在传输过程中是否遭篡改。

（2）加密机制：通过对数据进行加密运算来保证数据的机密性，以防数据在传输过程中被窃听。

（3）数据完整性：使 IP 通信的数据接收方能够确认数据在传输过程中是否遭篡改。

虽然 IPsec VPN 具有较高的安全性，但是也存在一些缺点。

（1）组网不灵活。建立 IPsec VPN，如果增加设备或调整用户的 IPsec 策略，需要调整原有 IPsec 配置。

（2）需要安装客户端软件，导致在兼容性、部署和维护方面都比较麻烦。

（3）只能基于五元组数据流建立 IPsec，无法明确区分出使用某终端接入 IPsec 的人是否为指定的授权用户，即管理员无法确知是谁在利用 VPN 访问内网资源。

某跨国公司的 IPsec VPN 防火墙配置实例如图 3-7 所示，该跨国公司的总公司、分公司以及出差到外地的远程用户均通过 IPsec VPN 实现安全通信。

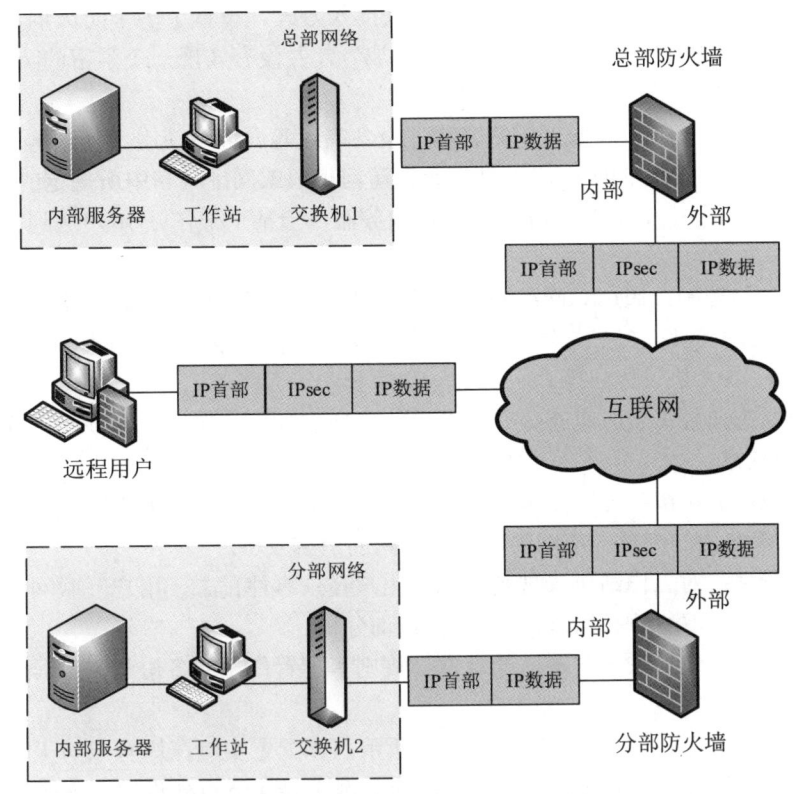

图 3-7 其跨国公司的 IPsec VPN 防火墙配置实例

3.4.3 分布式防火墙的部署

分布式防火墙是指驻留在网络主机并对主机系统提供安全防护的软件产品。分布式防火墙的基本思想是：安全策略由中心策略服务器集中定义，而安全策略的执行由相关主机节点独立实施；安全日志由主机节点分散产生，而安全日志集中保存在中心策略服务器上。

分布式防火墙负责网络边界、各子网和网络内部各节点之间的安全防护，根据其所需完成的功能，分布式防火墙的体系结构主要包括如下部分。

（1）网络防火墙。用于内部网络与外部网络之间以及内部网络各子网之间的防护。在功能上与传统的边界式防火墙类似，但与传统边界式防火墙相比，它多了一种用于内部子网之间的安全防护层，这样整个网络的安全防护体系就显得更加全面、更加可靠。

（2）主机防火墙。用于对网络中的服务器和个人计算机进行防护，实现应用层的安全防护，比网络层的安全防护更加彻底。这是传统边界式防火墙所不具有的，是对传统边界式防火墙在安全体系方面的一种完善。

（3）中心管理系统。这是分布式防火墙管理器软件，负责总体安全策略的制定、管理、分发以及日志的汇总，提高了防火墙的安全防护灵活性，同时具备高可管理性。

分布式防火墙的优点：首先，分布式防火墙一改先前的"信任内部、怀疑外部"的特点，实施了全方位的安全控制；其次，分布式防火墙一改先前的静态防火墙模式，实施了动态可扩展的结构体系；再次，分布式防火墙突破了先前的结构性瓶颈，实施了内部主机"步步为营"的战略部署；最后，分布式防火墙改变了 VPN 的接入模式，实施了简单的分布式接入。

分布式防火墙的不足：首先，分布式防火墙既要防内也要防外，压力很大，要求的技术及成本都很高；其次，分布式防火墙仅实现了多点接入方式，设置了多个防火墙，但其根本的防火墙管理并没有得到很好的实现，都是"单一作战"，并没有实现一个紧密的防火墙带。

分布式防火墙的主要应用包括以下几个方面。

（1）互联网访问控制：依据工作站名称、设备指纹等属性，使用"互联网访问规则"，控制工作站或工作站组在指定的时间段内是否允许访问模板或网址列表中所规定的互联网 Web 服务器，某个用户可否基于某工作站访问 www 服务器，当某个工作站/用户达到规定流量后是否断网。

（2）应用访问控制：通过对网络通信从链路层、网络层、传输层、应用层基于源地址、目的地址、端口、协议的逐层包过滤与入侵监测，控制来自局域网/互联网的应用服务请求，如 SQL 数据库访问、IPX 协议访问等。

（3）网络状态监控：实时动态报告当前网络中所有的用户登录、互联网访问、内网访问、网络入侵事件等信息。

（4）黑客攻击的防御：可以抵御包括 Smurf 拒绝服务攻击、ARP 欺骗式攻击、Ping 攻击、Trojan 木马攻击等在内的来自内部网络以及互联网的黑客攻击。

（5）日志管理：对工作站协议规则日志、用户登录事件日志、用户互联网访问日志、指纹验证规则日志、入侵检测规则日志进行记录与查询分析。

（6）系统工具：包括系统层参数的设定、规则等配置信息的备份与恢复、流量统计、模板设置、工作站管理等。

某高校的分布式防火墙配置实例如图 3-8 所示，该配置实例在原有双防火墙网络的基础上增加了分布式防火墙的安全体系结构，在网络的关键交界和节点处设置了安全屏障，从而形成了一个多层次、多协议、全方位安全的网络架构。

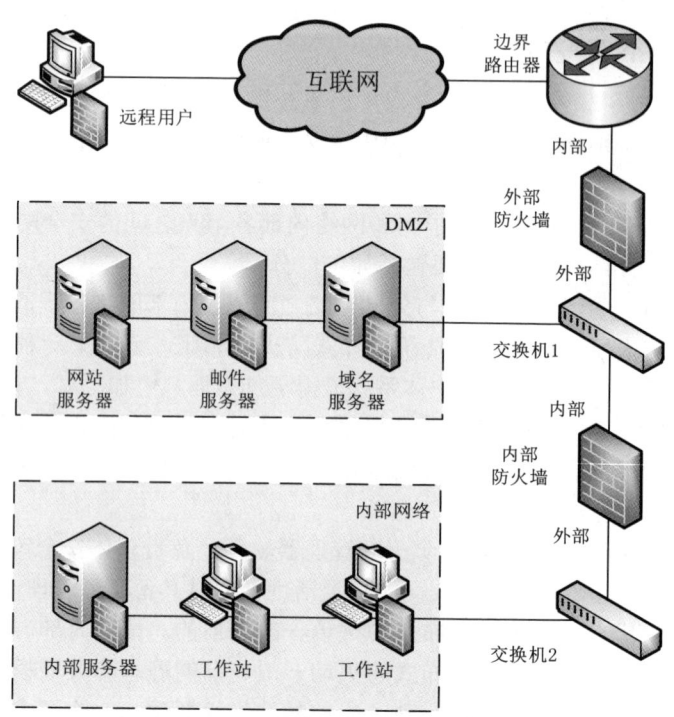

图 3-8　某高校的分布式防火墙配置实例

3.5 包过滤防火墙的配置

包过滤（packet filter）是所有防火墙最核心的功能，包过滤的标准是根据安全策略制定的，通常情况下由网络管理员在防火墙设备的 ACL 中设定。与代理服务器相比，它的优势是不占用网络带宽来传输信息。包过滤规则一般存放于路由器的 ACL 中，ACL 定义了各种规则来表明同意或拒绝数据包通过。如果没有一条规则能匹配，防火墙就会使用默认规则，一般情况下，默认规则要求防火墙丢弃该包。包过滤的核心是安全策略（包过滤算法）的设计。

3.5.1 ACL 概念

ACL（Access Control List，访问控制列表）是由 permit|deny 语句组成的一系列有顺序的规则，若干条规则构成一个访问控制列表。规则可以基于源地址、目的地址、源端口或目的端口等协议信息进行定义。路由器根据这些规则判断允许哪些数据包通过、拒绝哪些数据包通过，从而达到访问控制的目的。

不同厂商往往把 ACL 分为不同的类型，目前国内主要有两种主流的 ACL 系列：与 Cisco 兼容的系列（Cisco 系列）和与华为兼容的系列（华为系列）。Cisco 系列的 ACL 可分为标准 ACL 和扩展 ACL，其中标准 ACL 只能基于源 IP 地址定义规则，而扩展 ACL 可以基于源 IP 地址、目的 IP 地址、上层协议及其信息（如源端口号、目的端口号、ICMP 类型）定义规则，如表 3-1 所示。

表 3-1 ACL 规则定义条件

规则定义条件	标准 ACL	扩展 ACL
源 IP 地址	✓	✓
目的 IP 地址	×	✓
上层协议	×	✓
协议信息	×	✓

如何辨别 ACL 属于标准 ACL 还是扩展 ACL 呢？这可以在定义 ACL 时通过编号或关键字进行标识。对于编号范围及关键字的定义，不同的厂商往往有不同的方法。比如，Cisco 系列规定 ACL 类型与编号、关键字之间的对应关系如下。

- ➢ 编号
  - ◎ 标准 ACL：编号为 1～99 或 1300～1999；
  - ◎ 扩展 ACL：编号为 100～199 或 2000～2699。
- ➢ 关键字
  - ◎ 标准 ACL：关键字为 standard；
  - ◎ 扩展 ACL：关键字为 extended。

华为系列的 ACL 可分为基本 ACL、高级 ACL、接口 ACL 和 MAC ACL，其中基本 ACL、高级 ACL 与 Cisco 系列的标准 ACL 和扩展 ACL 类似。华为系列 ACL 的类型与编号、关键字之间的对应关系如下。

- ➢ 编号
  - ◎ 基本 ACL：编号为 2000～2999；

- ◎ 高级 ACL：编号为 3000～3999。
- ◎ 接口 ACL：编号为 1000～1999；
- ◎ MAC ACL：编号为 4000～4999。

➢ 关键字
- ◎ 基本 ACL：关键字为 basic；
- ◎ 高级 ACL：关键字为 advanced；
- ◎ 接口 ACL：关键字为 interface；
- ◎ MAC ACL：关键字为 MAC。

3.5.2 ACL 配置命令

路由器自顶向下逐一处理 ACL 语句，一旦发现匹配的语句，就不再处理列表中后面的语句。因此，语句的排列顺序显得尤为重要，ACL 语句排列的基本原则是：把最特殊的语句排在列表的最前面，而最一般的语句排在列表的最后面。注意：在 ACL 的最后一定存在一条缺省规则。

ACL 缺省规则：ACL 的最后总存在一条缺省规则，该规则是系统自动添加的。Cisco 系列的缺省规则默认为拒绝：deny any any；华为系列的缺省规则默认为允许：rule permit ip source any destination any。缺省规则的默认值是可以修改的，如将默认值由允许修改为拒绝：firewall default deny。

配置基于编号的标准 ACL 的命令：

```
R1(config)#access-list  ACL_#  {permit | deny}  conditions
```

标准 ACL 命令的完整格式：

```
Router(config)# access-list {1-99|1300-1999 } {permit 1 deny}  source_IP_address
[wildcard_mask]  [log]
```

ACL 匹配条件的定义：

```
source_IP_address [wildcard_mask]
```

其中，wildcard_mask（反子网掩码，32 位）规定："0"表示检查该位（相同才匹配）；"1"表示忽略该位（任意均匹配）。注意两个特别的反子网掩码，一个是全"0"：0.0.0.0，表示 ACL 语句中的 32 位地址要求全部匹配，因而叫作主机掩码，如 192.168.1.1 0.0.0.0 与 host 192.168.1.1 等价；另一个是全"1"：255.255.255.255，表示任意地址都是匹配的，通常与地址 0.0.0.0 一起使用，如 0.0.0.0 255.255.255.255 与 any 等价。

如果忽略反子网掩码，则默认为 0.0.0.0，即要求整个地址全部匹配。

下面给出几种典型 ACL 匹配条件的定义。

1. 匹配单台主机

例如，若要匹配一台 IP 地址为 1.1.1.1 的主机，则以下三种定义是等效的：

```
1.1.1.1   0.0.0.0
1.1.1.1                  //省略反子网掩码
host  1.1.1.1            //使用关键字"host"
```

2. 匹配任意主机

0.0.0.0 255.255.255.255 或使用关键字"any"。

3. 匹配一个连续的 IP 地址范围（区间）

1）匹配整个子网

例如，若某子网的 IP 地址为 1.1.3.16~1.1.3.31，其子网掩码为 255.255.255.240，则对应的反子网掩码可以通过下面的式子计算得到：

255.255.255.255−255.255.255.240=0.0.0.15

也可以用子网的末地址减去首地址得到：

1.1.3.31−1.1.3.16=0.0.0.15

故上述地址范围的匹配条件为 1.1.3.16 0.0.0.15。

2）匹配非完整子网范围

例如，若要匹配的 IP 地址为 1.1.1.31~1.1.1.67，由于该地址范围不是一个完整的子网范围，因此可以先将其分段，每段对应一个子网，用一条规则表示。上述 IP 地址可分为 3 段：1.1.1.31、1.1.1.32~1.1.1.63 和 1.1.1.64~1.1.1.67，故匹配条件分别为：

host 1.1.1.31

1.1.1.32 0.0.0.31

1.1.1.64 0.0.0.3

4. 匹配不连续的 IP 地址范围

若要匹配不连续的 IP 地址范围，则反子网掩码中各字节的值必然不等于 2^n-1。例如，匹配 192.168.10.1 0.0.0.254 条件的 IP 地址是 192.168.10.1～192.168.10.255 内的奇数 IP 地址。类似地，匹配 192.168.10.2 0.0.0.254 条件的 IP 地址是 192.168.10.1 至 192.168.10.255 内的偶数 IP 地址。若匹配条件是 3.2.9.128   0.0.0.20，则只有 4 个 IP 地址匹配：3.2.9.128、3.2.9.132、3.2.9.134 和 3.2.9.138。

ACL 的匹配过程（判断匹配与否）如下。

（1）用 ACL 中的反子网掩码和地址执行逻辑或。

（2）用 ACL 中的反子网掩码和接收到的分组首部中的 IP 地址执行逻辑或。

（3）将两个结果相减。若结果为零，则匹配；若结果不为零，则不匹配。

3.5.3 ACL 的应用

ACL 的典型应用可以分为两大类：一类用于数据包控制（包过滤），包括入栈应用（in）、出栈应用（out）；另一类用于诸如 NAT、IPsec 等控制。

当需要实现包过滤时，则在接口配置模式下应用 ACL，其配置命令为：

```
Router(config-if)#  ip  access-group  {ACL_# | ACL_ name }  { in 1 out}
```

注意：一个接口的"in"和"out"方向只能各应用一个 ACL。

在路由器接口 in 方向上应用 ACL（入栈应用）时，路由器的内部处理过程如图 3-9 所示。当 ACL 拒绝分组时，在默认的情况下，路由器会丢掉分组且向源主机发出 ICMP 出错报文。当然，出于安全或性能考虑，也可以配置路由器只是简单丢掉分组而不发出 ICMP 出错报文。

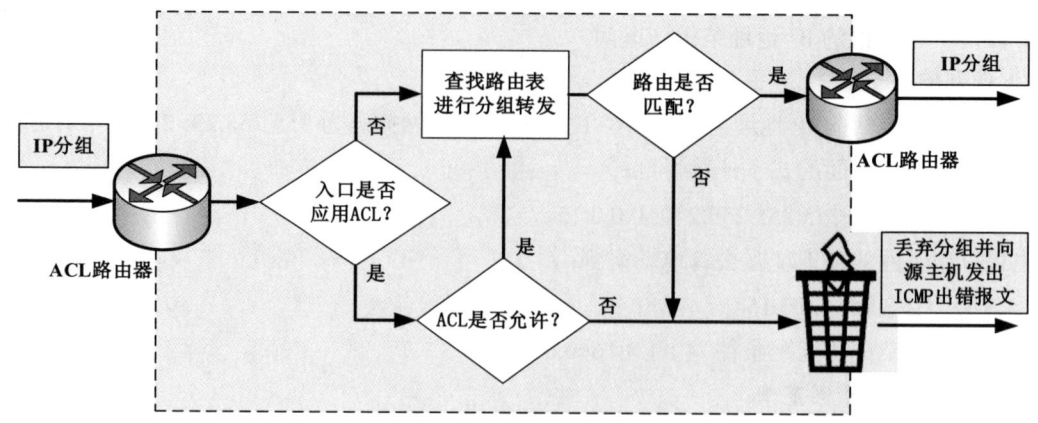

图 3-9 入栈应用处理过程

在路由器接口 out 方向上应用 ACL（出栈应用）时，路由器的内部处理过程如图 3-10 所示。当 ACL 拒绝分组时，在默认的情况下，路由器会丢掉分组且向源主机发出 ICMP 出错报文。当然，为了安全或性能考虑，也可以配置路由器只是简单丢掉分组而不发出 ICMP 出错报文。

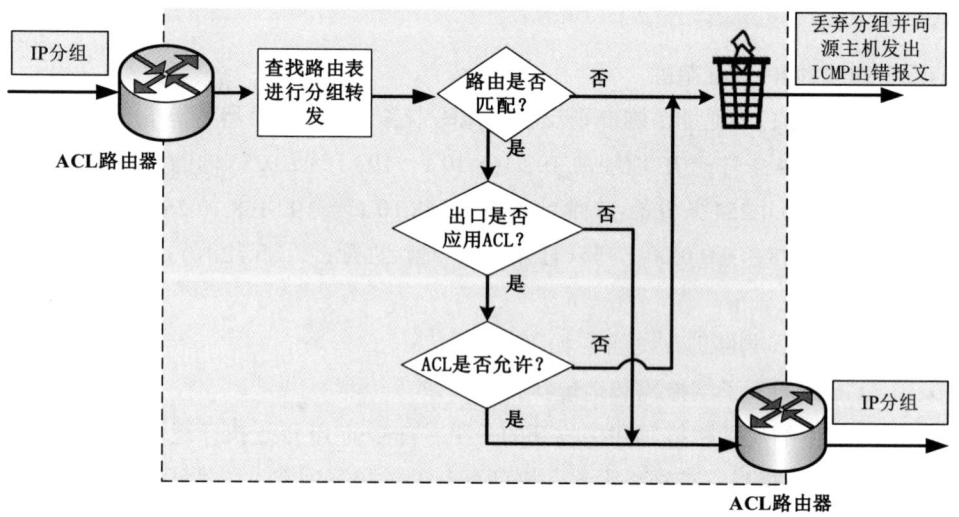

图 3-10 出栈应用处理过程

第 2 篇　交换机安全实践篇

第 4 章　安全远程登录网络设备

第 5 章　MAC 地址表及端口安全

第 6 章　多出口主机网关安全

第 7 章　IPv6 邻居发现协议

第 4 章
安全远程登录网络设备

4.1 组网需求

在实际的计算机网络中往往有大量的网络设备需要管理与维护，网络管理员总是通过网络设备的配置口 console 连接到设备中进行管理，通常是不可行的。此时，可以采用远程配置的方式，比如使用 Telnet 命令从当前设备登录网络上的另一台设备，从而实现对网络设备的远程管理与维护。但是，传统的 Telnet 命令只支持简单的密码认证，而不支持更安全的认证方式，而且 Telnet 在整个传输过程中不采用加密技术而是直接通过明文传输，存在比较大的安全隐患。

STelnet 是一种安全的 Telnet 服务，建立在 SSH（Secure Shell，安全外壳）连接的基础上。SSH 可以利用加密和强大的认证功能提供安全保障，保护设备不受 IP 地址欺诈、简单密码截取等攻击。

如图 4-1 所示，最左边的交换机作为 SSH 服务器，而最右边的 PC1 作为 SSH 客户端。SSH 客户端 PC1 可以通过密码、RSA、ECC 等多种认证方式登录 SSH 服务器，本实验以 RSA 认证方式说明客户端 PC1 是如何通过 STelnet 安全登录 SSH 服务器的。同时，为了进一步提升安全性，SSH 服务器上配置了 ACL 规则，用来防止非授权的用户非法登录 SSH 服务器。

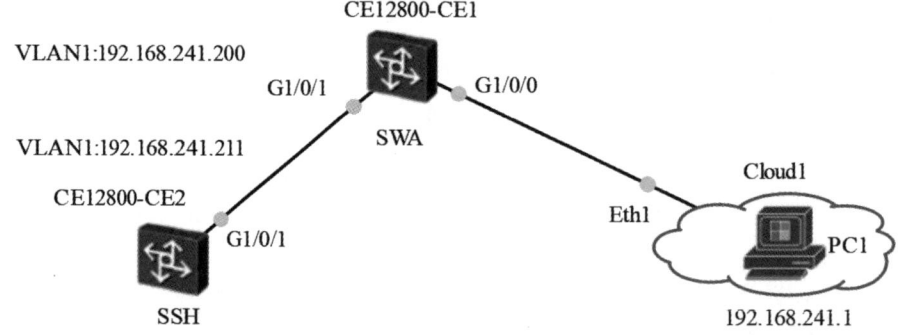

图 4-1 STelnet 网络拓扑

4.2 仿真实验

4.2.1 仿真环境

仿真环境所需要的硬件、软件建议至少满足以下要求。

- 物理机（PC）的处理器主频为 1.6GHz；
- 内存 RAM 大小为 32.0GB；
- 物理机的操作系统为 Windows 10；
- 虚拟机为 VMware Workstation 16 Pro；
- RSA 密钥生成软件为 PuTTYGen；
- 安全远程登录软件为 PuTTY；
- 网络模拟器选用 GNS3 2.2.31 版本；
- 交换机为华为 CE12800 系列交换机。

STelnet 网络拓扑如图 4-1 所示，两台交换机选用华为 CE12800 系列交换机，其中 CE12800-CE1（拓扑图中的设备名为 SWA，后续配置命令中的设备名为 cxySWA）经过接口 G1/0/0 与作为 SSH 客户端的 PC1 相连，经过接口 G1/0/1 与另一台作为 SSH 服务器的交换机 CE12800-CE2（拓扑图中的设备名为 SSH，后续配置命令中的设备名为 cxySSH）相连。

4.2.2　实验设计

在图 4-1 中，物理机（拓扑图中的 PC1）通过 Cloud1 的 Eth1 接口接入 GNS3 网络中，即对应虚拟网卡的 VMware Network Adapter VMnet8 网络。为了简单，两台交换机的 VLAN1 接口的 IP 地址和 VMnet8 网络的地址属于同一网络，即网络地址均为 192.168.241.0/24，这样 PC1 在网络层不用通过网关即可直接与 SSH 服务器通信。

按照前面的组网需求，希望在 PC1 上通过远程登录软件安全登录 SSH 服务器，具体的实现思路如下。

- 在网络层上客户端可以访问服务器，即在 PC1 上可以 ping 通 SSH 服务器 192.168.241.211；
- SSH 服务器开启 SSH、STelnet 服务，并创建对应的用户名和密码；
- 在 SSH 服务器上设置 ACL 允许客户端接入；
- 在物理机上通过密钥生成软件 PuTTYGen 生成 RSA 算法所需要的公钥和私钥对；
- 在物理机上借助远程登录软件 PuTTY 安全登录 SSH 服务器。

为此，在本实验操作前，需要完成下面的准备工作。

1. 安装并设置虚拟机

在物理机上安装虚拟机 VMware Workstation 16 Pro，具体安装过程在此省略。成功安装后，设置虚拟网卡 VMware Network Adapter VMnet8 的网络地址为 192.168.241.0/24，虚拟网卡 VMware Network Adapter VMnet1 的网络地址为 192.168.148.0/24，如图 4-2 所示。注意，此时 PC1 的 IP 地址默认值为 192.168.241.1/24，网关的 IP 地址默认值为 192.168.241.2/24。

2. 安装网络模拟器

在物理机上安装网络模拟器 GNS3 客户端（版本 2.2.31 或以上），接着在 VMware Workstation 16 Pro 中安装 GNS3 服务器虚拟机，具体过程省略。完成后，启动 GNS3 客户端并在"Setup Wizard"窗口中选择"Run appliances on a remote server (advanced usage)"单选按钮（如图 4-3 所示），然后单击"Next"按钮。

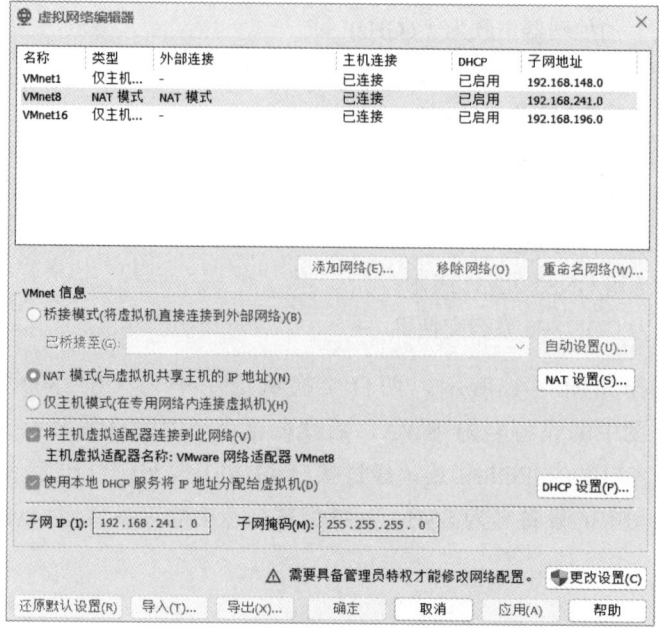

图 4-2　设置虚拟网卡的网络地址

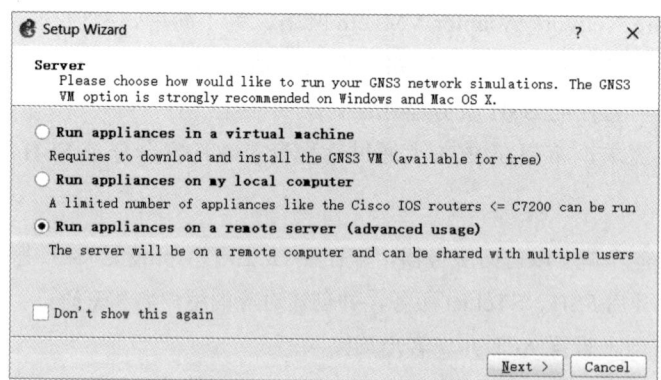

图 4-3　选择 GNS3 服务器

接着，在弹出窗口的"Host"文本框中输入 GNS3 服务器的 IP 地址"192.168.148.132"，如图 4-4 所示，然后单击"Next"按钮。

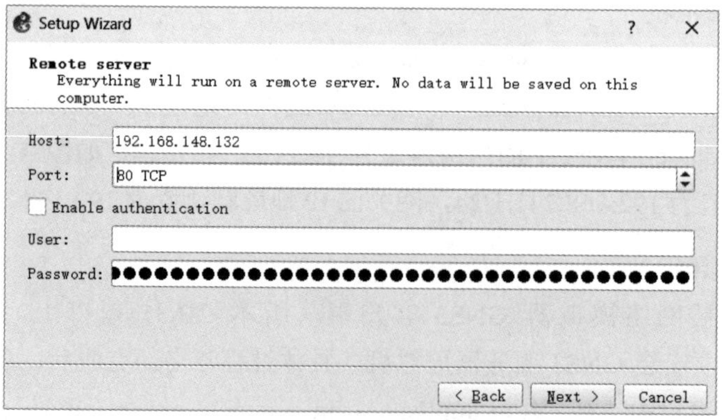

图 4-4　设置 GNS3 服务器的 IP 地址

最后，在弹出的窗口（如图4-5所示）中单击"Finish"按钮，完成GNS3服务器的设置。

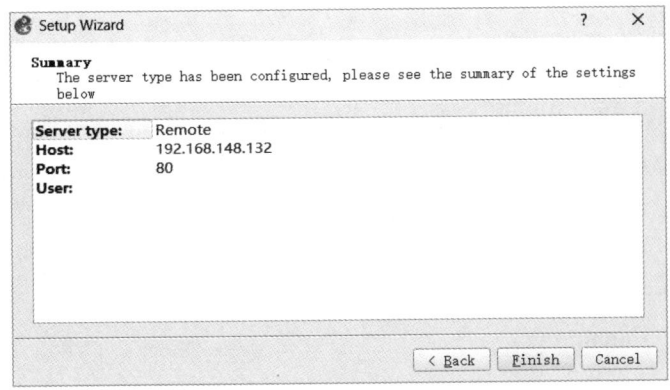

图4-5　GNS3服务器设置成功

3. 安装密钥生成和安全远程登录软件

密钥生成和安全远程登录软件的安装很简单，只需在物理机中完成下列操作即可：将密钥生成软件对应的可执行文件"PuTTYGen.exe"和安全远程登录软件对应的可执行文件"PuTTY.exe"复制到D盘根目录下。

4.2.3　操作步骤

下面给出本次实验的具体操作，一共包括13个步骤。

1. 按照图4-1所示的网络拓扑，在GNS3中新建一个空白的工程

两台交换机选用华为的CE12800系列二层交换机，版本为"Version 8.180"，即具体型号为"HuaWei CE12800 V200R005C10SPC607B607"。

物理机（拓扑图中的PC1）通过Cloud的Eth1接口（对应虚拟网卡的VMware Network Adapter VMnet8网络，IP地址为192.168.241.1/24）接入GNS3网络中。

2. 将SWA重命名为cxySWA，并配置SWA交换机的VLAN1接口IP地址

```
<HUAWEI> system-view imm
[HUAWEI] sysname cxySWA
[cxySWA] interface vlan1
[cxySWA-Vlanif1] ip address 192.168.241.200 24
        // 192.168.241.*为虚拟网卡VMnet8的网络地址，必要时请将之修改为实际值
```

3. 将SSH重命名为cxySSH，配置SSH服务器的VLAN1接口IP地址

```
<HUAWEI> system-view imm
[HUAWEI] sysname cxySSH
[cxySSH] interface vlan1
[cxySSH-Vlanif1] ip address 192.168.241.211 24
        // 192.168.241.*为虚拟网卡VMnet8的网络地址，必要时请将之修改为实际值
[cxySSH-Vlanif1] quit
```

4. 配置 cxySSH 服务器的 VTY 用户界面，包括认证方式、特权等级

[cxySSH] user-interface vty 0 4

[cxySSH-ui-vty0-4] authentication-mode aaa

Warning: The level of the user-interface(s) will be the default level of AAA users, please check whether it is correct. After the authentication mode is set to AAA, you need to enter the user name and password to log in.

[cxySSH-ui-vty0-4] protocol inbound ssh

[cxySSH-ui-vty0-4] user privilege level 3

[cxySSH-ui-vty0-4] quit

5. 在 cxySSH 服务器上创建本地用户"sshuser-1"，密码为"Cxy@12345"，将用户加入管理员组，并指明用户的服务类型为"SSH"

[cxySSH] aaa

[cxySSH-aaa] local-user sshuser-1 password irreversible-cipher Cxy@12345

Info: A new user is added.

[cxySSH-aaa] local-user sshuser-1 user-group manage-ug   // 用户的组名为 manage-ug

[cxySSH-aaa] local-user sshuser-1 service-type ssh

[cxySSH-aaa] quit

6. 配置服务器的源接口为 VLAN1

[cxySSH]ssh server-source -i Vlanif 1

Warning: SSH server source configuration will take effect in the next login. Do you want to continue? [Y/N]:y

Info: Succeeded in setting the source interface of the SSH server to Vlanif1.

7. 在服务器端创建 SSH 用户"sshuser-1"，并配置认证方式为 RSA

[cxySSH] ssh user sshuser-1

Info: Succeeded in adding a new SSH user.

[cxySSH] ssh user sshuser-1 authentication-type rsa

8. SSH 客户端 PC1 使用 PuTTYGen 创建 RSA 密钥对，并将公钥复制至 SSH 服务器

（1）打开 PuTTYGen 软件，如图 4-6 所示。在该图的"Parameters"选区中选择"RSA"单选按钮，然后在"Number of bits in a generated key"文本框中输入"2048"，最后单击"Generate"按钮。

（2）在产生密钥对的过程中，需要在程序窗口的"Key"选区的空白区域内不停地移动鼠标，如图 4-7 所示，直至密钥对成功生成。

（3）成功生成密钥对后，首先，在图 4-8 所示的"Key"选区的"Key passphrase"文本框中输入密码（注意：此密码将作为 SSH 用户登录 SSH 服务器的密码，因此需要记住。假设将密码设置为"cxy"），输入完毕后在"Confirm passphrase"文本框中再次输入上述密码"cxy"。

其次，在"Actions"选区中单击"Save private key"按钮，弹出如图 4-9 所示的"Save private key as"对话框，在该对话框中输入私钥文件名（如"cxyRSAprivate-lab1"）用于保存私钥，然后单击"保存"按钮。最后，在图 4-8 所示的"Key"选区中将生成的公钥全部复制并粘贴至字处理软件（如记事本或写字板）中，并以文本类型保存，文件可以命名为"RSApublic.txt"。

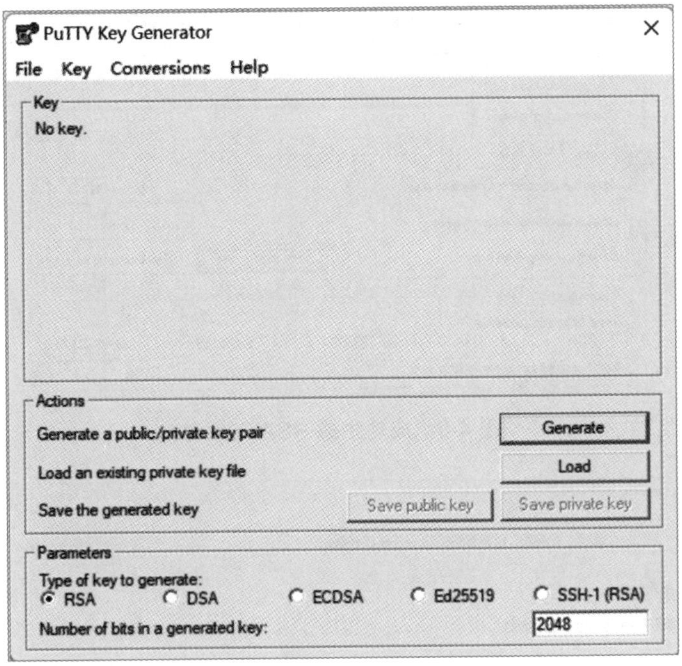

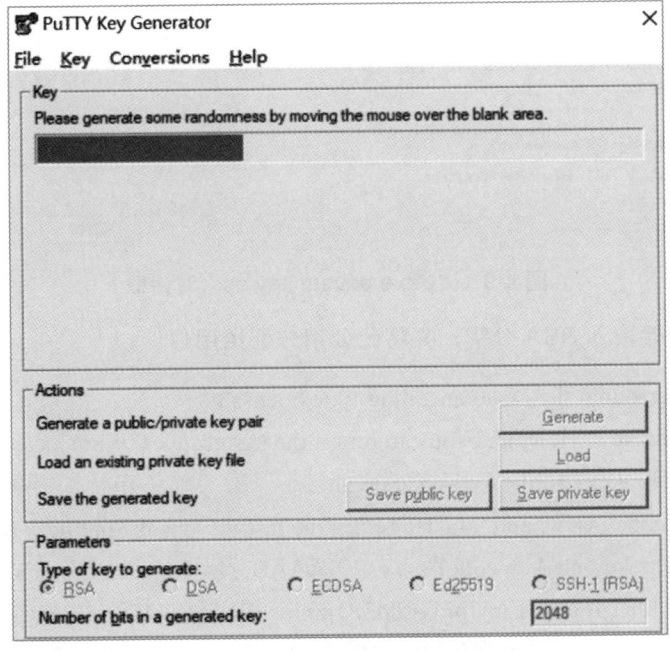

图 4-6　准备创建 RSA 密钥对

图 4-7　正在生成 RSA 密钥对

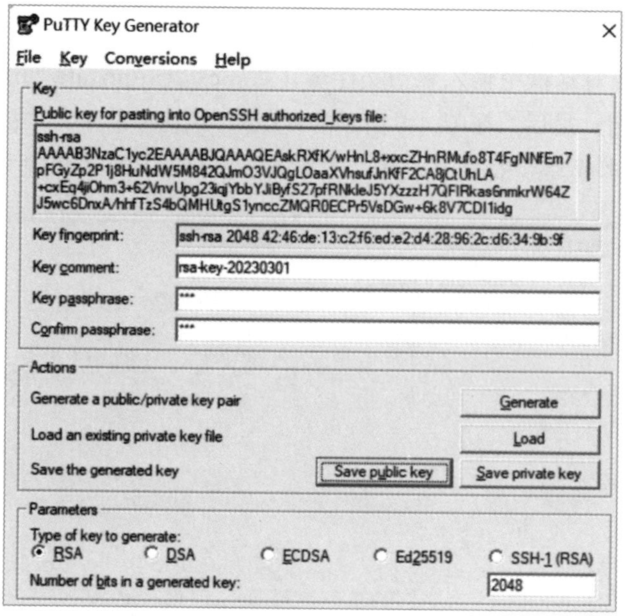

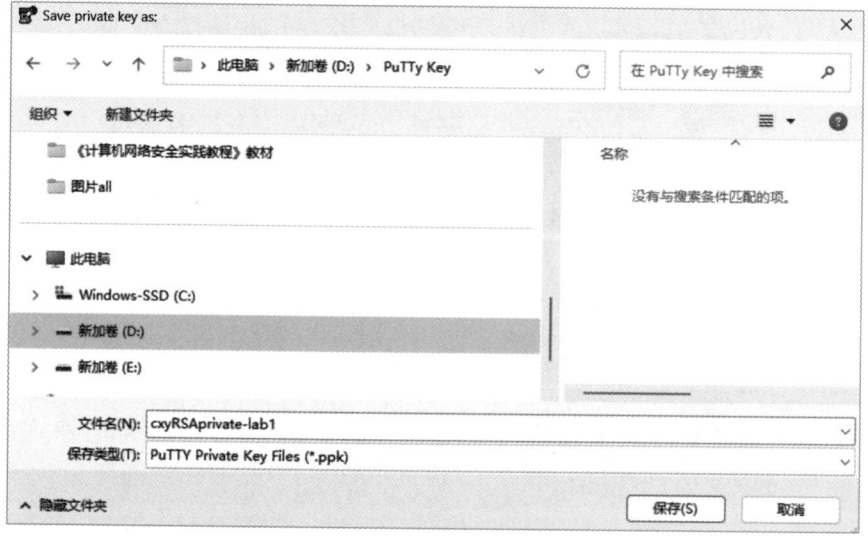

图 4-8　成功生成 RSA 密钥对

图 4-9　"Save private key as"对话框

9. 向 SSH 服务器输入 RSA 公钥，并将此公钥分配给用户

[cxySSH] rsa peer-public-key rsa0 encoding-type openssh
Enter "RSA public key" view, and you can return the system view with "peer-public-key end".
[cxySSH-rsa-public-key] public-key-code begin
Enter "RSA key code" view, and you can return the last view with "public-key-code end".
[cxySSH-rsa-public-key-rsa-key-code] ssh-rsa AAAAB3NzaC1yc2EAAAABJQAAAQEAskRXfK/wHnL8+xxcZHnRMufo8T4FgNNfEm7pFGyZp2P1j8HuNdW5M842QJmO3VJQgLOaaXVhsufJnKfF2CA8jCtUhLA+cxEq4jiOhm3+62VnvUpg23iqjYbbYJiByfS27pfRNkleJ5YXzzzH7QFIRkas6nmkrW64ZJ5wc6DnxA/hhfTzS4bQMHUtgS1ynccZMQR0ECPr5VsDGw+6k8V7CDI1idg+LoNEypUzfDaMx

Z7r/NSZdnLnGNUk2j5sC6ya64ALy+P3U7E8jZO7aRitd50paHyFaixBvs99jCo3zFgQd7/YwhJtLN8h
LAo4GmHgZmJFgk18CCvQbHHfVsVT1Q== rsa-key-20230301
// 粘贴刚才复制到剪贴板的公钥（SSH 客户端的公钥）
[cxySSH-rsa-public-key-rsa-key-code] public-key-code end
[cxySSH-rsa-public-key] peer-public-key end
[cxySSH] ssh user sshuser-1 assign rsa-key rsa0

10. 启用 STelnet 功能，并创建服务类型为 STelnet 的用户"sshuser-1"

[cxySSH]stelnet server enable
Info: Succeeded in starting the STelnet server.
[cxySSH]ssh user sshuser-1 service-type stelnet

11. 配置 ACL 规则，允许源 IP 地址为 192.168.241.0/29 的主机访问

[cxySSH] acl 2000
[cxySSH-acl4-basic-2000] rule permit source 192.168.241.0 0.0.0.7
[cxySSH-acl4-basic-2000] quit
[cxySSH] ssh server acl 2000

12. 客户端 PC1 通过 PuTTY 软件登录 SSH 服务器

（1）在 PC1 中打开 PuTTY 软件，即在物理机中以管理员身份运行"cmd"命令，在弹出的"命令提示符"窗口中输入"putty"命令运行 SSH 客户端程序，如图 4-10 所示。接着，将 SSH 服务器的 IP 地址"192.168.241.211"输入"Host Name (or IP address)"文本框中。

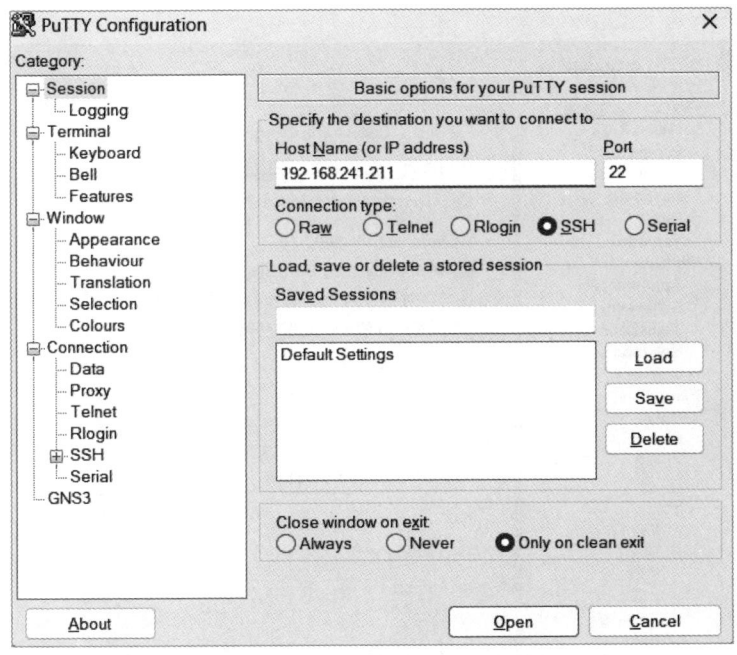

图 4-10　通过 PuTTY 软件登录 SSH 服务器（1）

（2）在图 4-11 中，单击"Category"→"Connection"→"SSH"选项，在"Protocol

options"选区中,选择"2 only"单选按钮。

(3)在图 4-11 中,单击"SSH"→"Auth"选项,出现图 4-12 所示的界面,单击"Authentication parameters"选区中的"Browse"按钮,导入前面保存的私钥文件 cxyRSAprivate-lab1.ppk,然后单击"Open"按钮。

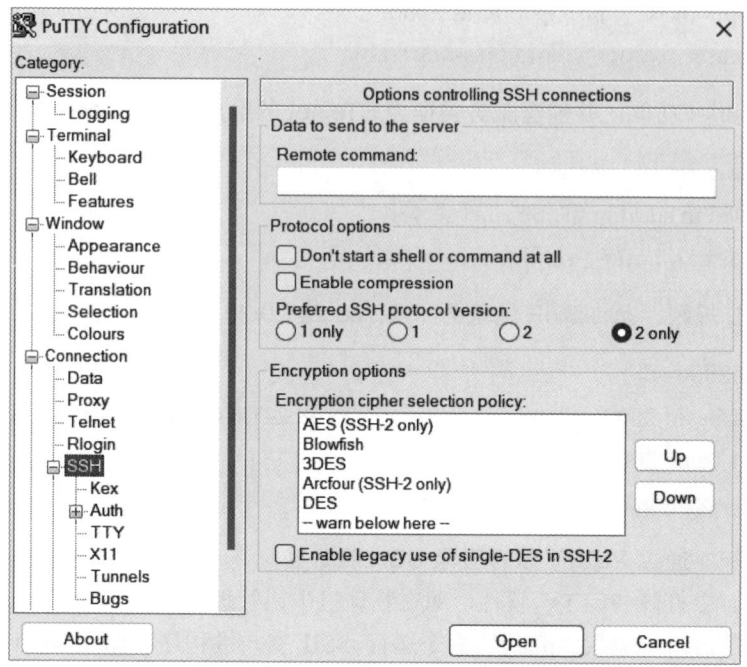

图 4-11　通过 PuTTY 软件登录 SSH 服务器(2)

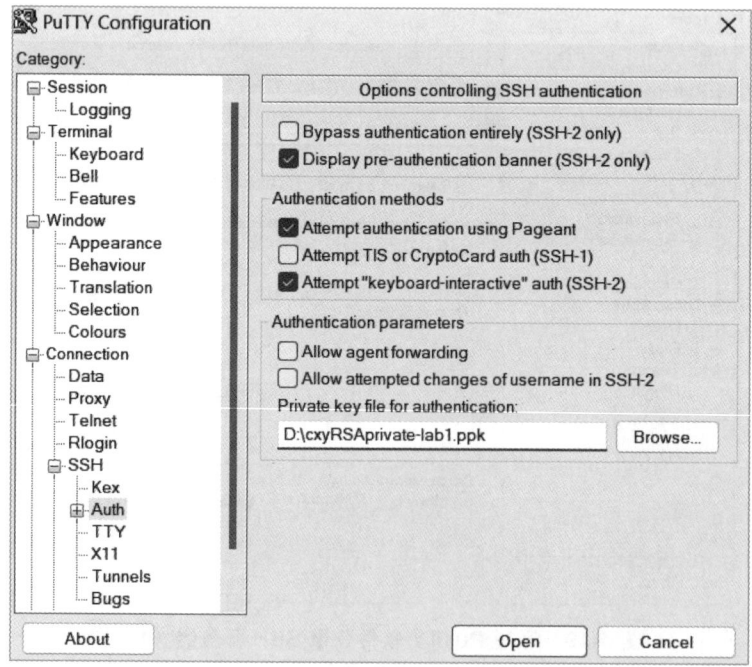

图 4-12　通过 PuTTY 软件登录 SSH 服务器(3)

13. 在弹出的窗口中，输入用户名（sshuser-1）和密码（cxy），可成功地安全远程登录 SSH 服务器（如图 4-13 所示）

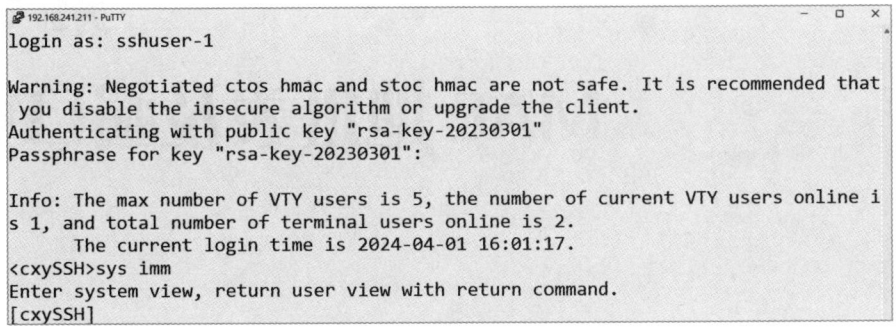

图 4-13 安全远程登录 SSH 服务器

第 5 章
MAC 地址表及端口安全

5.1 组网需求

在如图 5-1 所示的网络拓扑中，PC1 至 PC4 均属于同一个 VLAN，为防止非法用户假冒别人的 MAC 地址窃取重要用户信息，需要在交换机 SW1 上配置静态 MAC 地址绑定，未绑定的 MAC 地址禁止接入网络。

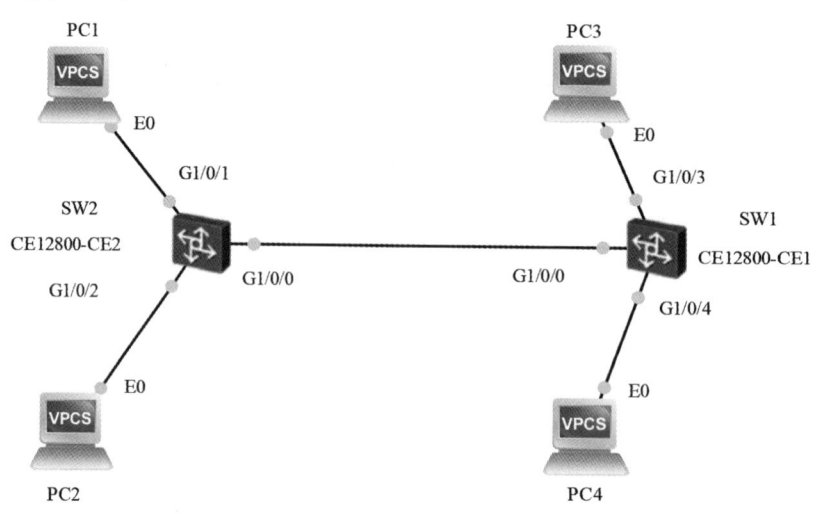

图 5-1　网络拓扑

此外，为了防止黑客通过 MAC 地址攻击用户设备或网络，要求在交换机 SW1 上实现过滤非法 MAC 地址的功能，即通过配置黑洞 MAC 表项，将黑客的 MAC 地址配置为黑洞 MAC 地址。当设备收到目的 MAC 或源 MAC 地址为黑洞 MAC 地址的报文，且报文携带的 VLAN ID 为黑洞 MAC 对应的 VLAN 时，直接丢弃。

在此基础上，还要求交换机 SW1 具有端口安全功能，且端口安全的类型可以是以下三类中的任意一种：基于安全动态 MAC 地址的端口安全、基于安全静态 MAC 地址的端口安全和基于 Sticky-MAC 地址的端口安全。

安全动态 MAC 地址指的是启用端口安全而未启用 Sticky-MAC 功能时转换的 MAC 地址，其特点是设备重启后，安全动态 MAC 地址表项就会丢失，需要重新学习。在默认情况下，安全动态 MAC 地址不会老化，只有在配置安全 MAC 老化时间后才会老化。

安全静态 MAC 地址指的是启用端口安全时在交换机上手工配置的静态 MAC 地址，其特点是不会老化，只要配置保存，即使设备重启也不会丢失。

Sticky-MAC 地址指的是交换机先进行安全动态 MAC 地址的学习，然后再将学习到的 MAC 地址保存到 MAC 地址表项中，不需要通过手工配置，可以防止手动配置错误且可以自动更新。

对于违反上述任意一条安全策略的，可以在交换机 SW1 上启用以下三种安全端口保护动作：Restrict、Protect 和 Shutdown。其中，Restrict 表示丢弃源 MAC 地址违反安全策略的帧，同时发出告警信息。Protect 表示丢弃源 MAC 地址违反安全策略的帧，但不发出告警信息。Shutdown 表示端口将被置为 error-down 状态（关闭该端口），同时发出告警信息。在默认情况下，端口关闭后不会自动恢复，只能由网络管理员通过一定的命令重启才能恢复正常状态。

5.2 仿真实验

5.2.1 仿真环境

仿真环境所需要的硬件、软件建议至少满足以下要求。

- 物理机的处理器主频为 1.6GHz；
- 内存 RAM 大小为 32.0GB；
- 物理机的操作系统为 Windows 10；
- 虚拟机为 VMware Workstation 16 Pro；
- 网络模拟器选用 GNS3 2.2.31；
- 交换机为华为 CE12800；
- PC1~PC4 为 Virtual PC Simulator, Version 0.8.2。

5.2.2 实验设计

根据网络拓扑（图 5-1）完成网络设备的选型：两台交换机选用华为的 CE12800 系列二层交换机，其中 CE12800-CE1（拓扑图中的设备名为 SW1，后续配置命令中的设备名为 cxySW1）经过接口 G1/0/3 和 PC3 相连、经过接口 G1/0/4 和 PC4 相连、经过接口 G1/0/0 与另一台交换机 CE12800-CE2（拓扑图中的设备名为 SW2，后续配置命令中的设备名为 cxySW2）相连。根据前面的组网需求，本实验设计如下。

1. VLAN 设计

在交换机 SW1 中创建 VLAN2，并将接口 G1/0/0、G1/0/3、G1/0/4 和 G1/0/9 划分到 VLAN2 中；在交换机 SW2 中创建 VLAN2，并将接口 G1/0/0、G1/0/1 和 G1/0/2 划分到 VLAN2 中。

2. 静态 MAC 地址绑定设计

配置 SW1 的静态 MAC 地址表项，将 PC1、PC2、PC3 和 PC4 的 MAC 地址分别绑定到 VLAN2 中的以下接口：G1/0/0、G1/0/0、G1/0/3 和 G1/0/4。

在 VLAN2 上配置 MAC 地址学习限制规则，最多可以学习 2 个 MAC 地址，对超过最大 MAC 地址学习数量的报文进行告警提示并丢弃。

3. 端口安全功能设计

配置黑洞 MAC 表项，把 PC1 的 MAC 地址配置为黑洞 MAC 地址。

配置 G1/0/0 的端口安全功能，启用端口 Sticky MAC 功能，配置端口安全功能的保护动作为"error-down"，配置端口 MAC 地址学习数量不超过 5 个。

5.2.3 操作步骤

下面给出本实验的具体操作，一共包括 8 个步骤。

1. 按照图 5-1 所示的网络拓扑，在 GNS3 中新建一个空白的工程

两台交换机选用华为的 CE12800 系列二层交换机，版本为"Version 8.180"，即具体型号为"HuaWei CE12800 V200R005C10SPC607B607"。

4 台 PC 选择 GNS3 提供的虚拟机 VPCS 即可。

2. 配置 PC 的 IP 地址，并记下接口的 MAC 地址

```
PC1> ip 2.0.0.1/8
Checking for duplicate address...
PC1: 2.0.0.1 255.0.0.0

PC1>show ip
NAME:            PC1[1]
IP/MASK          :        2.0.0.1/8
GATEWAY          :        0.0.0.0
DNS              :
MAC              :        00:50:79:66:68:01
LPORT            :        20032
RHOST:PORT       :        127.0.0.1:20033
MTU              :        1500
```

记下 PC1 的 MAC 地址：00:50:79:66:68:01。同理，设置 PC2～PC4 的 IP 地址依次为：2.0.0.2/8、2.0.0.3/8 和 2.0.0.4/8，PC2～PC4 的 MAC 地址依次为：00:50:79:66:68:02、00:50:79:66:68:03 和 00:50:79:66:68:00。

3. 在交换机 SW1 中创建 VLAN2，并将接口 G1/0/0、G1/0/3、G1/0/4 和 G1/0/9 划分到 VLAN2 中

```
[HUAWEI] sysname cxySW1          // 可将 cxy 改为你的名字拼音首字母
[cxySW1] vlan2
[cxySW1-vlan2] interface g1/0/0
[cxySW1-GE1/0/0] port link-type trunk
[cxySW1-GE1/0/0] port trunk allow-pass vlan2
[cxySW1-GE1/0/0] undo port trunk allow-pass vlan1
[cxySW1-GE1/0/0] quit
```

```
[cxySW1] interface g1/0/3
[cxySW1-GE1/0/3] port link-type access
[cxySW1-GE1/0/3] port default vlan2
[cxySW1-GE1/0/3] quit
[cxySW1] interface g1/0/4
[cxySW1-GE1/0/4] port link-type access
[cxySW1-GE1/0/4] port default vlan2
[cxySW1-GE1/0/4] quit
[cxySW1] interface g1/0/9
[cxySW1-GE1/0/9] port link-type access
[cxySW1-GE1/0/9] port default vlan2
```

4. 在交换机 SW2 中创建 VLAN2,并将接口 G1/0/0、G1/0/1 和 G1/0/2 划分到 VLAN2 中

```
[HUAWEI] sysname cxySW2      // 可将 cxy 改为你的名字拼音首字母
[cxySW2] vlan2
[cxySW2-vlan2] interface g1/0/0
[cxySW2-GE1/0/0] port link-type trunk
[cxySW2-GE1/0/0] port trunk allow-pass vlan2
[cxySW2-GE1/0/0] undo port trunk allow-pass vlan1
[cxySW2-GE1/0/0] quit
[cxySW2] interface g1/0/1
[cxySW2-GE1/0/1] port link-type access
[cxySW2-GE1/0/1] port default vlan2
[cxySW2-GE1/0/1] quit
[cxySW2] interface g1/0/2
[cxySW2-GE1/0/2] port link-type access
[cxySW2-GE1/0/2] port default vlan2
[cxySW2-GE1/0/2] quit
```

至此,PC1~PC4 4 台 PC 能够互通,可以通过 ping 命令验证。在交换机 cxySW1 上用 display mac-address 显示 MAC 地址表。

```
[cxySW1] display mac-address
Flags: *Backup
BD: bridge-domain   Age: dynamic MAC learned time in seconds
MAC Address       VLAN/VSI/BD    Learned-From    Type       Age
0050-7966-6800    2/-/-          GE1/0/4         dynamic    -
0050-7966-6801    2/-/           GE1/0/0         dynamic    -
0050-7966-6802    2/-/           GE1/0/0         dynamic    -
0050-7966-6803    2/-/           GE1/0/3         dynamic    -

Total items: 4
```

请问若将 PC4 从接口 G1/0/3 移到接口 G1/0/9，此时 PC4 是否可以 ping 通其他 PC？为什么？

5. 配置 cxySW1 的静态 MAC 地址表项，分别将 PC1、PC2、PC3 和 PC4 的 MAC 地址绑定到 VLAN2 中的以下接口：G1/0/0、G1/0/0、G1/0/3 和 G1/0/4

```
[cxySW1] mac-address static 0050-7966-6801 g1/0/0 vlan2
[cxySW1] mac-address static 0050-7966-6802 g1/0/0 vlan2
[cxySW1] mac-address static 0050-7966-6803 g1/0/3 vlan2
[cxySW1] mac-address static 0050-7966-6800 g1/0/4 vlan2
```

在交换机 cxySW1 上用 display mac-address 显示 MAC 地址表。我们会发现，此时的类型变成了"static"：

```
[cxySW1] display mac-address
Flags:  *Backup
BD:  bridge-domain  Age: dynamic MAC learned time in seconds
MAC Address      VLAN/VSI/BD    Learned-From      Type       Age
0050-7966-6800   2/-/-          GE1/0/4           static     -
0050-7966-6801   2/-/           GE1/0/0           static     -
0050-7966-6802   2/-/           GE1/0/0           static     -
0050-7966-6803   2/-/           GE1/0/3           static     -

Total items: 4
```

请问若将 PC4 从接口 G1/0/4 切换到接口 G1/0/9，再在 PC4 上 ping PC3，是否可以 ping 通？为什么？需要如何配置才能 ping 通？

6. 在 VLAN2 上配置 MAC 地址学习限制规则，最多可以学习 2 个 MAC 地址，对超过最大 MAC 地址学习数量的报文进行告警提示并丢弃（注意：限制地址学习的数量不包含静态的 MAC 地址）

```
[cxySW1] vlan2
[cxySW1-vlan2] mac-address limit maximum 2 alarm enable action discard
```

先用命令 undo mac-address static 撤销前面配置的静态 MAC 地址，然后在 PC4 上分别 ping 另外两台 PC——PC1 和 PC2：

```
PC4> ping 2.0.0.1

84 bytes from 2.0.0.1 icmp_seq=1 ttl=64 time=6.669 ms
84 bytes from 2.0.0.1 icmp_seq=2 ttl=64 time=12.926 ms
84 bytes from 2.0.0.1 icmp_seq=3 ttl=64 time=7.271 ms
84 bytes from 2.0.0.1 icmp_seq=4 ttl=64 time=9.921 ms
84 bytes from 2.0.0.1 icmp_seq=5 ttl=64 time=12.425 ms

PC4> ping 2.0.0.2
```

host (2.0.0.2) not reachable

可见，此时只能 ping 通第一台 PC，而 ping 不通后面的 PC，请分析其中的原因。请分析说明利用交换机的这个特性，是否可以在一定程度上防止拒绝服务攻击。

7. 配置黑洞 MAC 表项，把 PC1 的 MAC 地址配置为黑洞 MAC 地址

[cxySW1-vlan2] undo mac-address limit
[cxySW1-vlan2] quit
[cxySW1] mac-address blackhole 0050-7966-6801 vlan2

配置黑洞 MAC 地址后，在 PC1 上用 ping 命令进行测试：

PC1> ping 2.0.0.2

84 bytes from 2.0.0.2 icmp_seq=1 ttl=64 time=7.068 ms
84 bytes from 2.0.0.2 icmp_seq=2 ttl=64 time=2.320 ms
84 bytes from 2.0.0.2 icmp_seq=3 ttl=64 time=3.257 ms
84 bytes from 2.0.0.2 icmp_seq=4 ttl=64 time=6.055 ms
84 bytes from 2.0.0.2 icmp_seq=5 ttl=64 time=5.166 ms

PC1> ping 2.0.0.3

host (2.0.0.3) not reachable

PC1> ping 2.0.0.4

host (2.0.0.4) not reachable

可见，PC1 不能通过交换机 cxySW1 访问其他 PC，但是可以通过 cxySW2 交换机与 PC2 进行通信。

8. 配置 G1/0/0 的端口安全功能

// 启用端口安全功能
[cxySW1-GE1/0/0] port-security enable
// 启用端口 Sticky MAC 功能
[cxySW1-GE1/0/0] port-security mac-address sticky
// 配置端口安全功能的保护动作为"error-down"
[cxySW1-GE1/0/0] port-security protect-action error-down
// 配置端口 MAC 地址学习数量不超过 5 个
[cxySW1-GE1/0/0] port-security maximum 5
[cxySW1-GE1/0/0] quit

第 6 章 多出口主机网关安全

6.1 组网需求

某公司有两个互联网出口,其网络拓扑如图 6-1 所示,其中路由器 R1 和 R2 分别模拟 ISP1 和 ISP2 的接入网路由器,SW1、SW2、SW3 模拟公司内部网络的交换机。PC1、PC2 同属于 VLAN10,PC3、PC4 同属于 VLAN20;PC1、PC2 的 IP 地址分别设置为 10.10.0.11/8 和 10.10.0.22/8,PC3、PC4 的 IP 地址分别设置为 20.20.0.11/8 和 20.20.0.22/8。

现要求通过配置路由冗余协议,在提高公司网络内主机访问互联网可靠性的同时实现负载均衡。在图 6-1 中,把同一 VLAN 内的主机的网关设置成相同的,以实现可靠性的提高:将 VLAN10 内的 PC1 和 PC2 的网关均设置为备份组 10 的虚拟 IP 地址 10.0.0.1。

在此基础上,把同一 VLAN 内的主机的网关设置成不相同的,则可保证在可靠性提高的同时实现负载均衡:将 VLAN10 内的 PC1 的网关设置为备份组 10 的虚拟 IP 地址 10.0.0.1,而将 PC2 的网关设置为备份组 20 的虚拟 IP 地址 10.0.0.2。

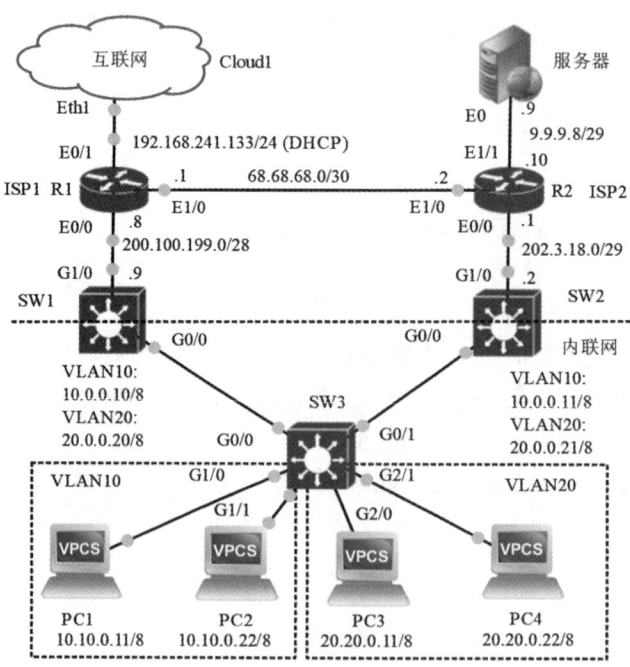

图 6-1 网络拓扑

6.2 仿真实验

6.2.1 仿真环境

仿真环境所需要的硬件、软件建议至少满足以下要求。

- 物理机（PC）的处理器主频为 1.6GHz；
- 内存 RAM 大小为 32.0GB；
- 物理机的操作系统为 Windows 10；
- 虚拟机为 VMware Workstation 16 Pro；
- 网络模拟器选用 GNS3 2.2.31；
- 路由器为 Cisco IOL Router 15.7；
- 交换机为 Cisco IOSvL2 2020；
- PC1～PC4 为 Virtual PC Simulator, Version 0.8.2。

6.2.2 实验设计

根据网络拓扑（图 6-1）完成网络设备的选型：3 台交换机选用 Cisco IOSvL2 2020 系列三层交换机，其中 SW1 经过接口 G0/0 和 SW3 的接口 G0/0 相连。两台路由器选用 Cisco IOL Router 15.7 系列路由器，其中 R1 经过接口 E1/0 和 R2 的接口 E1/0 相连。根据前面的组网需求，本实验设计如下。

1. VLAN 设计

在 3 台交换机 SW1、SW2 和 SW3 中分别创建两个 VLAN：VLAN10 和 VLAN20，并将各自的接口 G0/0 设置为 trunk 模式并允许所有 VLAN 通过；对于交换机 SW3，还需要将接口 G1/0 和 G1/1 划分到 VLAN10 中，将接口 G2/0 和 G2/1 划分到 VLAN20 中。

2. IP 地址规划

网络中共有 7 个不同的网段，网络地址分别为 200.100.199.0/28、192.168.241.0/24、68.68.68.0/30、202.3.18.0/29、9.9.9.8/29、10.0.0.0/8 和 20.0.0.0/8。各设备接口的 IP 地址/掩码如表 6-1 所示。

表 6-1 各设备接口的 IP 地址/掩码

设备	接口	IP 地址/掩码
R1	E0/0	200.100.199.8/28
R1	E0/1	192.168.241.133/24
R1	E1/0	68.68.68.1/30
R2	E0/0	202.3.18.1/29
R2	E1/0	68.68.68.2/30
R2	E1/1	9.9.9.10/29
SW1	VLAN10	10.0.0.10/8
SW1	VLAN20	20.0.0.20/8
SW2	VLAN10	10.0.0.11/8
SW2	VLAN20	20.0.0.21/8
PC1～PC4		见图 6-1

3. 网络路由设计

在两台三层交换机 SW1 和 SW2 上均开启路由功能，同时启用 OSPF 动态路由，在 SW1 和 SW2 的 3 个接口上均声明为 OSPF 主干网络。此外，在 SW1 上设置一条下一跳是 R1 接口 200.100.199.8 的默认路由，而在 SW2 上设置一条下一跳是 R2 接口 202.3.18.1 的默认路由。

在两台路由器 R1 和 R2 上启用 OSPF 动态路由，并在 R1 和 R2 的两个接口上均声明为 OSPF 主干网络。此外，在 R1 上设置一条到达目的网络 9.9.9.8/29 下一跳是 R2 接口 68.68.68.2 的静态路由；而在 R2 上设置一条下一跳是 R1 接口 68.68.68.1 的默认路由。

4. HSRP 设计

在 SW1 交换机 VLAN10 接口上设置备份组 10 和备份组 20，其中备份组 10 的虚拟 IP 地址是 10.0.0.1，优先级正常时是 150（当接口 G1/0 的链路状态变为 DOWN 时，优先级降为 60），允许抢占别人使自己成为主路由器；而备份组 20 的虚拟 IP 地址 10.0.0.2，优先级固定为 100，允许抢占别人使自己成为主路由器。

在 SW1 交换机 VLAN20 接口上设置备份组 10 和备份组 20，其中备份组 10 的虚拟 IP 地址是 20.0.0.1，优先级固定为 200，允许抢占；而备份组 20 的虚拟 IP 地址是 20.0.0.2，优先级正常是 100（当接口 G1/0 的链路状态变为 DOWN 时，优先级降为 50），允许抢占。

类似地，在 SW2 交换机 VLAN10 接口上设置备份组 10 和备份组 20，其中备份组 10 的虚拟 IP 地址是 10.0.0.1，优先级正常时是 100（当接口 G1/0 的链路状态变为 DOWN 时，优先级降为 50），允许抢占；而备份组 20 的虚拟 IP 地址是 10.0.0.2，优先级正常时是 150（当接口 G1/0 的链路状态变为 DOWN 时，优先级降为 50），允许抢占。

同理，在 SW2 交换机 VLAN20 接口上设置备份组 10 和备份组 20，其中备份组 10 的虚拟 IP 地址是 20.0.0.1，优先级正常时是 250（当接口 G1/0 的链路状态变为 DOWN 时，优先级降为 150），允许抢占；而备份组 20 的虚拟 IP 地址是 20.0.0.2，优先级固定为 90，允许抢占。

6.2.3 操作步骤

下面给出本次实验的具体操作，一共包括 9 个步骤。

1. 按照图 6-1 所示的网络拓扑，在 GNS3 中新建一个空白的工程

在 GNS3 中完成网络设备的选型及组网，其中，路由器 R1、R2 选择"Cisco IOL Router 15.7"，交换机 SW1~SW3 选择"Cisco IOSvL2 2020"，PC1~PC4 及服务器均选择 VPCS。在连接到 Cloud1 时，注意要连接到接口 Eth1，即连接到虚拟机的 NAT 网络。

2. 配置 3 台交换机 VLAN

1）在交换机 SW1 上创建 VLAN10 和 VLAN20，并把接口划分到 VLAN 中

```
Switch(config)#hostname cxySW1              // 可将 cxy 改为你的名字拼音首字母
cxySW1(config)#line console 0
cxySW1(config-line)#exec-timeout 300 0      // 设置空闲超时时长为 300 分钟 0 秒
cxySW1(config)#vlan10
cxySW1(config-vlan)#vlan20
cxySW1(config-vlan)#interface g0/0
```

```
cxySW1(config-if)#switchport trunk encapsulation dot1q
cxySW1(config-if)#switchport mode trunk
```

2）在交换机 SW2 上创建 VLAN10 和 VLAN20，并把接口划分到 VLAN 中

```
Switch(config)#hostname cxySW2        // 可将 cxy 改为你的名字拼音首字母
cxySW2(config)#vlan10
cxySW2(config-vlan)#vlan20
cxySW2(config-vlan)#interface g0/0
cxySW2(config-if)#switchport trunk encapsulation dot1q
cxySW2(config-if)#switchport mode trunk
```

3）在交换机 SW3 上创建 VLAN10 和 VLAN20，并把接口划分到 VLAN 中

```
Switch(config)#hostname cxySW3        // 可将 cxy 改为你的名字拼音首字母
cxySW3(config)#vlan10
cxySW3(config-vlan)#vlan20
cxySW3(config-vlan)#interface g0/0
cxySW3(config-if)#switchport trunk encapsulation dot1q
cxySW3(config-if)#switchport mode trunk
cxySW3(config-vlan)#interface g0/1
cxySW3(config-if)#switchport trunk encapsulation dot1q
cxySW3(config-if)#switchport mode trunk
cxySW3(config)#interface g1/0
cxySW3(config-if)#switchport access vlan10
cxySW3(config-if)#interface g1/1
cxySW3(config-if)#switchport access vlan10
cxySW3(config-if)#interface g2/0
cxySW3(config-if)#switchport access vlan20
cxySW3(config-if)#interface g2/1
cxySW3(config-if)#switchport access vlan20
```

3. 配置接口 IP 地址

1）配置交换机 SW1 的 VLAN10 和 VLAN20 接口 IP 地址

```
cxySW1(config)#interface g1/0
cxySW1(config-if)#no switchport
cxySW1(config-if)#ip address 200.100.199.9 255.255.255.240
cxySW1(config-if)#no shutdown
cxySW1(config)#interface vlan10
cxySW1(config-if)#ip address 10.0.0.10 255.0.0.0
cxySW1(config-if)#no shutdown
cxySW1(config)#interface vlan20
```

cxySW1(config-if)#ip address 20.0.0.20 255.0.0.0
cxySW1(config-if)#no shutdown

2）配置交换机 SW2 的 VLAN10 和 VLAN2 接口 IP 地址

cxySW2(config)# interface g1/0
cxySW2(config-if)#no switchport
cxySW2(config-if)#ip address 202.3.18.2 255.255.255.248
cxySW2(config-if)#no shutdown
cxySW2(config)#interface vlan10
cxySW2(config-if)#ip address 10.0.0.11 255.0.0.0
cxySW2(config-if)#no shutdown
cxySW2(config)#interface vlan20
cxySW2(config-if)#ip address 20.0.0.21 255.0.0.0
cxySW2(config-if)#no shutdown

3）配置路由器 R1 接口 IP 地址

R1(config)#hostname cxyR1
cxyR1(config)#interface e0/1
cxyR1(config-if)#ip address dhcp
cxyR1(config-if)#no shutdown
cxyR1(config-if)#interface e0/0
cxyR1(config-if)#ip address 200.100.199.8 255.255.255.240
cxyR1(config-if)#no shutdown
cxyR1(config)#interface e1/0
cxyR1(config-if)#ip address 68.68.68.1 255.255.255.248
cxyR1(config-if)#no shutdown

4）配置路由器 R2 接口 IP 地址

R2(config)#hostname cxyR2
cxyR2(config)#interface e1/0
cxyR2(config-if)#ip address 68.68.68.2 255.255.255.248
cxyR2(config-if)#no shutdown
cxyR2(config-if)#interface e0/0
cxyR2(config-if)#ip address 202.3.18.1 255.255.255.248
cxyR2(config-if)#no shutdown
cxyR2(config)#interface e1/1
cxyR2(config-if)#ip address 9.9.9.10 255.255.255.248
cxyR2(config-if)#no shutdown

4. 配置路由，使得所有 IP 地址在网络层互通

1）配置 SW1 交换机的静态路由和 OSPF 动态路由

cxySW1(config)#ip routing　　　　　　　　//开启三层交换机的路由功能

```
cxySW1(config)#ip route 0.0.0.0 0.0.0.0 200.100.199.8
cxySW1(config)#router ospf 1
cxySW1(config-router)#network 10.0.0.0 0.0.0.255 area 0
cxySW1(config-router)#network 20.0.0.0 0.0.0.255 area 0
cxySW1(config-router)#network 200.100.199.0 0.0.0.15 area 0
```

2）配置 SW2 交换机的默认路由和 OSPF 动态路由

```
cxySW2(config)#ip routing
cxySW2(config)#ip route 0.0.0.0 0.0.0.0 202.3.18.1
cxySW2(config)#router ospf 1
cxySW2(config-router)#network 10.0.0.0 0.0.0.255 area 0
cxySW2(config-router)#network 20.0.0.0 0.0.0.255 area 0
cxySW2(config-router)#network 202.3.18.0 0.0.0.7 area 0
```

3）配置路由器 R1 的静态路由和 OSPF 动态路由

```
cxyR1(config)#ip route 9.9.9.8 255.255.255.248 68.68.68.2
cxyR1(config)#router ospf 1
cxyR1(config-router)#network 200.100.199.0 0.0.0.15 area 0
cxyR1(config-router)#network 68.68.68.0 0.0.0.3 area 0
```

4）配置路由器 R2 的默认路由和 OSPF 动态路由

```
cxyR2(config)#ip route 0.0.0.0 0.0.0.0 68.68.68.1
cxyR2(config)#router ospf 1
cxyR2(config-router)#network 68.68.68.0 0.0.0.3 area 0
cxyR2(config-router)#network 202.3.18.0 0.0.0.7 area 0
```

5. 配置 PC 和服务器的 IP 地址、网关和 DNS 服务器

```
PC1> ip 10.10.0.11/8 10.0.0.10
PC1> ip dns 8.8.8.8
PC2> ip 10.10.0.22/8 10.0.0.10
PC2> ip dns 8.8.8.8
PC3> ip 20.20.0.11/8 20.0.0.21
PC3> ip dns 8.8.8.8
PC4> ip 20.20.0.22/8 20.0.0.21
PC4> ip dns 8.8.8.8
Server> ip 9.9.9.9/29 9.9.9.10
```

6. 跟踪内网主机访问互联网的路由

1）从 PC1 上用 trace 命令跟踪到达 www.phei.com.cn 的路由

```
PC1> trace www.phei.com.cn
trace to www.phei.com.cn,  8 hops max,  press Ctrl+C to stop
```

1	10.0.0.10	4.699 ms	2.778 ms	6.955 ms
2	200.100.199.8	6.765 ms	6.729 ms	3.285 ms
3	192.168.241.2	5.103 ms	3.240 ms	4.080 ms
4	*	*	*	

2）从 PC3 上用 trace 命令跟踪到达 www.phei.com.cn 的路由

```
PC3>trace www.phei.com.cn
trace to www.phei.com.cn, 8 hops max, press Ctrl+C to stop
 1  20.0.0.21      6.593ms    8.859ms    6.794ms
 2  202.3.18.1     8.669ms    6.894ms    8.272ms
 3  68.68.68.1     7.013ms    6.684ms    7.828ms
 4  192.168.241.2  8.065ms    5.762ms    6.329ms
 5  *              *          *
```

PC3> ping www.phei.com.cn -t
cxySW2(config)#interface g1/0
cxySW2(config-if)#shutdown   // 模拟 ISP2 出口出现故障
PC3>ping www.phei.com.cn -t
84 bytes from 163.177.151.109 icmp_seq=1 ttl=126 time=21.478ms
84 bytes from 163.177.151.109 icmp_seq=2 ttl=126 time=25.056ms
84 bytes from 163.177.151.109 icmp_seq=3 ttl=126 time=21.772ms
　*20.0.0.21 icmp_seq=4 ttl=255 time=6.672ms (ICMP type:3, code: 1, Destination host unreachable)
　*20.0.0.21 icmp_seq=5 ttl=255 time=4.952ms (ICMP type:3, code: 1, Destination host unreachable)
　*20.0.0.21 icmp_seq=6 ttl=255 time=4.886ms (ICMP type:3, code: 1, Destination host unreachable)

可见，此时PC3、PC4不能上网。注意，此时PC1和PC2是可以上网的。
思考：如何在ISP2出口出现故障时，令PC3和PC4可以上网呢？
手工修改PC3和PC4的网关试试：

PC3> ip 20.20.0.11/8 20.0.0.20

PC4> ip 20.20.0.22/8 20.0.0.20

此时，PC3、PC4可以ping通www.phei.com.cn吗？它们的出口是哪个ISP？
思考：如果不手工修改PC的网关，在ISP2出口出现故障时能否实现上网呢？
测试完成后，恢复SW2接口G1/0为正常状态：

cxySW2(config)#interface g1/0

cxySW2(config-if)#no shutdown

7. HSRP 设置

1）在 SW1 交换机 VLAN10 接口上设置备份组 10 和备份组 20

```
cxySW1(config)#track 1 interface g1/0 line-protocol
#定义名为 1 的跟踪对象，该对象跟踪接口 G1/0 的链路状态
cxySW1(config)#interface vlan10
cxySW1(config-if)#standby 10 ip 10.0.0.1          // 备份组 10 的虚拟 IP 地址 10.0.0.1
cxySW1(config-if)#standby 10 priority 150         // 优先级为 150
cxySW1(config-if)#standby 10 preempt              // 允许抢占别人使自己成为主路由器
cxySW1(config-if)#standby 10 track 1 decrement 90
// 引用跟踪对象 1，使得当接口 G1/0 的链路状态变为 DOWN 时，优先级降为 90

cxySW1(config-if)#standby 20 ip 10.0.0.2          // 备份组 20 的虚拟 IP 地址 10.0.0.2
cxySW1(config-if)#standby 20 preempt              // 允许抢占
// 未设置 "priority"，故优先级为默认值 100
```

注意：备份组 ID 的作用域是广播域内（本子网内），故下面 VLAN20 内的备份组 10、20 跟前面 VLAN10 内的备份组 10、20 无任何关联。

因此，在判别不同路由器的接口是否属于同一个备份组时，可以基于备份组的虚拟 IP 地址进行（若基于备份组 ID，则还需考虑备份组的作用域是否属于同一个广播域）。若两台或更多台路由器的备份组虚拟 IP 地址相同，则它们属于同一个备份组。

2）在 SW1 交换机 VLAN20 接口上设置备份组 10 和备份组 20

```
cxySW1(config)#interface vlan20
cxySW1(config-if)#standby 10 ip 20.0.0.1          // 备份组 10 的虚拟 IP 地址 20.0.0.1
cxySW1(config-if)#standby 10 priority 200         // 优先级为 200
cxySW1(config-if)#standby 10 preempt              // 允许抢占
#未设置跟踪接口 "track"，故优先级不会改变

cxySW1(config-if)#standby 20 ip 20.0.0.2          // 备份组 20 的虚拟 IP 地址 20.0.0.2
cxySW1(config-if)#standby 20 preempt              // 允许抢占
cxySW1(config-if)#standby 20 track 1 decrement 50
// 引用跟踪对象 1，使得当接口 G1/0 的链路状态变为 DOWN 时，优先级降为 50
// 未设置 "priority"，故优先级为默认值 100
// 若未设置 "preempt"，则不允许抢占（本路由器优先级高于主路由器时，也不能为主路由器）
```

3）在 SW2 交换机 VLAN10 接口上设置备份组 10 和备份组 20

```
cxySW2(config)#track 1 interface g1/0 line-protocol
// 定义名为 1 的跟踪对象，该对象跟踪接口 G1/0 的链路状态
cxySW2(config)#interface vlan10
```

```
cxySW2(config-if)#standby 10 ip 10.0.0.1
cxySW2(config-if)#standby 10 track 1 decrement 50
cxySW2(config-if)#standby 10 preempt

cxySW2(config-if)#standby 20 ip 10.0.0.2
cxySW2(config-if)#standby 20 priority 150
cxySW2(config-if)#standby 20 track 1 decrement 100
cxySW2(config-if)#standby 20 preempt
```

4)在 SW2 交换机 VLAN20 接口上设置备份组 10 和备份组 20

```
cxySW2(config-if)#interface vlan20
cxySW2(config-if)#standby 10 ip 20.0.0.1
cxySW2(config-if)#standby 10 priority 250
cxySW2(config-if)#standby 10 preempt
cxySW2(config-if)#standby 10 track 1 decrement 100
cxySW2(config-if)#standby 20 ip 20.0.0.2
cxySW2(config-if)#standby 20 priority 90
cxySW2(config-if)#standby 20 preempt
```

8. 查看备份组相关信息

1)查看 SW1 交换机 VLAN10 接口的备份组的详细信息

```
cxySW1#show standby vlan10
Vlan10-Group 10
State is Active
2 state changes. last state change 08:10:43
Virtual IP address is 10.0.0.1
Active virtual MAC address is 0000.0c07.ac0a (MAC In Use)
Local virtual MAC address is 0000.0c07.ac0a (v1 default)
Hello time 3 sec, hold time 10 sec
Next hello sent in 0.016 sec
Preemption enabled
Active router is local
Standby router is 10.0.0.11, priority 100 (expires in 10.656 sec)
Priority 150 (configured 150)
Track object 1 state Up decrement 90
Group name is"hsrp-Vl10-10" (default)
Vlan10-Group 20
State is Standby
4 state changes. last state change 03:06:06
Virtual IP address is 10.0.0.2
```

```
Active virtual MAC address is 0000.0c07.ac14 (MAC Not In Use)
Local virtual MAC address is 0000.0c07.ac14 (v1 default)
Hello time 3 sec, hold time 10 sec
Next hello sent in 2.080 sec
Preemption enabled
Active router is 10.0.0.11, priority 150 (expires in 10.128 sec)
Standby router is local
Priority 100 (default 100)
Group name is "hsrp-Vl10-20" (default)
```

2）查看 SW1 交换机路由冗余备份组的摘要信息

```
cxySW1#show standby brief
                P indicates configured to preempt.
```

Interface	Grp	Pri	P	State	Active	Standby	Virtual IP
Vl10	10	150	P	Active	local	10.0.0.11	10.0.0.1
Vl10	20	100	P	Standby	10.0.0.11	local	10.0.0.2
Vl20	10	200	P	Standby	20.0.0.21	local	20.0.0.1
Vl20	20	100		Active	local	20.0.0.21	20.0.0.2

可见，在 VLAN10 的备份组 10 中：主路由器（Active）是 SW1，而 SW2 是备份路由器（Standby）；而在备份组 20 中：主路由器（Active）是 SW2，而 SW1 是备份路由器（Standby）。

9. 路由冗余调试

1）网络正常时，测试 PC1 和 PC2 访问互联网的路由

（1）设置 PC1 和 PC2 的网关为 VLAN10 的备份组 10 的虚拟 IP 地址 10.0.0.1。

```
PC1> ip 10.10.0.11/8 10.0.0.1
PC2> ip 10.10.0.22/8 10.0.0.1
```

（2）分别在 PC1 和 PC2 上进行路由跟踪。

```
PC1> trace www.phei.com.cn
trace to www.phei.com.cn, 8 hops max,    press Ctrl+C to stop
 1   10.0.0.10           7.819ms        6.815ms        7.786ms
 2   200.100.199.8       4.903ms        3.501ms        4.206ms
 3   192.168.241.2       4.918ms        6.994ms        8.221ms
 4   *          *          *
PC2> trace www.phei.com.cn
trace to www.baidu.com, 8 hops max,    press Ctrl+C to stop
 1   10.0.0.10           3.440ms        3.595ms        5.331ms
 2   200.100.199.8       4.937ms        2.742ms        3.924ms
 3   192.168.241.2       4.242ms        5.535ms        2.984ms
 4   *          *          *
```

请分析此时 PC1 和 PC2 访问互联网，是通过哪个 ISP 进行的？

2）SW1 接口 G1/0 出现故障时，通过路由冗余协议可提高可靠性

（1）模拟 ISP1 出口出现故障。

```
cxySW1(config)#interface g1/0
cxySW1(config-if)#shutdown
```

（2）查看交换机 SW1 备份组的状态。

```
cxySW1#show standby brief
                     P indicates configured to preempt.
Interface   Grp   Pri    P   State     Active     Standby   Virtual IP
Vl10        10    60     P   Standby   10.0.0.11  local     10.0.0.1
Vl10        20    100    P   Standby   10.0.0.11  local     10.0.0.2
Vl20        10    200    P   Standby   20.0.0.21  local     20.0.0.1
Vl20        20    50     P   Standby   20.0.0.21  local     20.0.0.2
```

（3）查看交换机 SW2 备份组的状态。

```
cxySW2#show standby brief
                     P indicates configured to preempt.
Interface   Grp   Pri    P   State     Active   Standby    Virtual IP
Vl10        10    100    P   Active    local    10.0.0.10  10.0.0.1
Vl10        20    150    P   Active    local    10.0.0.10  10.0.0.2
Vl20        10    250    P   Active    local    20.0.0.20  20.0.0.1
Vl20        20    90     P   Active    local    20.0.0.20  20.0.0.2
```

（4）从 PC1 上跟踪访问互联网的路由。

```
PC1> trace www.phei.com.cn
trace to www.phei.com.cn, 8 hops max,    press Ctrl+C to stop
 1   10.0.0.11        7.511ms      7.712ms      10.616ms
 2   202.3.18.1       9.632ms      7.326ms      6.749ms
 3   68.68.68.1       8.997ms      7.693ms      7.044ms
 4   192.168.241.2    11.502ms     10.483ms     5.325ms
 5   *        *         *
```

可见，PC1 可以访问互联网。

思考：此时的出口是哪个 ISP？路由冗余协议 HSRP 的作用是什么？能否说明将同一 VLAN 内的主机的网关设置成相同的（均设置为备份组 10 的虚拟 IP 地址 10.0.0.1），则可提高可靠性？

```
cxySW1(config)#interface g1/0
cxySW1(config-if)#no shutdown                    // ISP1 出口故障排除
```

3）网络正常时，通过路由冗余协议可实现负载均衡

实现负载均衡的基本思想是将同一 VLAN 内的主机的网关设置成不相同的：将 PC1 的网关

设置为备份组 10 的虚拟 IP 地址 10.0.0.1，而将 PC2 的网关设置为备份组 20 的虚拟 IP 地址 10.0.0.2。

（1）设置 PC1 和 PC2 的网关分别为 VLAN10 的备份组 10 的虚拟 IP 地址 10.0.0.1 和备份组 20 的虚拟 IP 地址 10.0.0.2。

```
PC1> ip 10.10.0.11/8 10.0.0.1
PC2> ip 10.10.0.22/8 10.0.0.2
```

（2）在 PC1 上进行路由跟踪。

```
PC1> trace www.phei.com.cn
trace to www.phei.com.cn, 8 hops max，press Ctrl+C to stop
1   10.0.0.10        4.431ms       6.174ms       5.292ms
2   200.100.199.8    4.441ms       8.826ms       12.862ms
3   192.168.241.2    11.150ms      12.216ms      9.447ms
4   *       *        *
```

（3）在 PC2 上进行路由跟踪。

```
PC2> trace www.phei.com.cn
trace to www.phei.com.cn, 8 hops max，press Ctrl+C to stop
1   10.0.0.11        13.490ms      9.269ms       8.695ms
2   202.3.18.1       10.331ms      12.283ms      10.407ms
3   68.68.68.1       11.612ms      10.892ms      12.259ms
4   192.168.241.2    16.016ms      12.314ms      14.253ms
5   *       *        *
```

可见，当网络正常时，PC1 和 PC2 分别通过 ISP1 和 ISP2 访问互联网，实现了负载均衡。

4）SW1 接口 G1/0 出现故障时，通过路由冗余协议可提高可靠性

（1）模拟 ISP1 出口出现故障。

```
cxySW1(config)#interface g1/0
cxySW1(config-if)#shutdown            // 模拟 ISP1 出口出现故障
```

（2）在 PC1 上进行路由跟踪。

```
PC1> trace www.phei.com.cn
trace to www.phei.com.cn, 8 hops max，press Ctrl+C to stop
1   10.0.0.11        6.231ms       3.904ms       10.601ms
2   202.3.18.1       8.459ms       15.802ms      9.374ms
3   68.68.68.1       8.760ms       8.189ms       8.106ms
4   192.168.241.2    9.587ms       7.818ms       6.918ms
5   *       *        *
```

（3）在 PC2 上进行路由跟踪。

```
PC2> trace www.phei.com.cn
trace to www.phei.com.cn, 8 hops max, press Ctrl+C to stop
   1  10.0.0.11         13.490ms        9.269ms         8.695ms
   2  202.3.18.1        10.331ms        12.283ms        10.407ms
   3  68.68.68.1        11.612ms        10.892ms        12.259ms
   4  192.168.241.2     16.016ms        12.314ms        14.253ms
   5  *        *        *
```

可见，当 ISP1 出现故障时，PC1 和 PC2 均通过 ISP2 访问互联网。

```
cxySW1(config)#interface g1/0
cxySW1(config-if)#no shutdown                    // 模拟 ISP1 出口故障排除
```

5）SW2 接口 G1/0 出现故障时，通过路由冗余协议可提高可靠性

（1）模拟 ISP2 出口出现故障。

```
cxySW2(config)#interface g1/0
cxySW2(config-if)#no shutdown
```

（2）在 PC1 上进行路由跟踪。

```
PC1> trace www.phei.com.cn
trace to www.phei.com.cn, 8 hops max, press Ctrl+C to stop
   1  10.0.0.10         7.392ms         10.075ms        8.716ms
   2  200.100.199.8     8.702ms         11.139ms        6.394ms
   3  192.168.241.2     10.246ms        9.611ms         11.823ms
   4  *        *        *
```

（3）在 PC2 上进行路由跟踪。

```
PC2> trace www.phei.com.cn
trace to www.phei.com.cn, 8 hops max, press Ctrl+C to stop
   1  10.0.0.10         8.030ms         7.543ms         9.793ms
   2  200.100.199.8     8.599ms         11.059ms        6.123ms
   3  192.168.241.2     8.043ms         9.020ms         6.114ms
   4  *        *        *
```

可见，当 ISP2 出现故障时，PC1 和 PC2 均通过 ISP1 访问互联网。

```
cxySW2(config)#interface g1/0
cxySW2(config-if)#no shutdown                    // 模拟 ISP2 出口故障排除
```

类似地，对 VLAN20 内的 PC3 和 PC4 设置合适的网关，也可实现在保证负载均衡的同时提高可靠性。请自行配置并加以验证。

第 7 章
IPv6 邻居发现协议

7.1 组网需求

如图 7-1 所示，两台交换机通过 G1/0/0 相连，它们的 G1/0/1、G1/0/2 均属于 VLAN100。为 VLAN100、VLAN200 接口配置 IPv6 链路本地地址、全球单播地址和任播地址，实现相应主机间的互通。

RFC4861 中定义了邻居发现协议（Neighbor Discovery Protocol，NDP），通过 NDP 协议可以实现 MAC 地址解析、重复地址检测（Duplicate Address Detect，DAD）、邻居状态跟踪等功能。

下面以主机 A 解析主机 B 的 MAC 地址为例说明地址解析的过程。主机 A 在向主机 B 发送报文之前必须解析出主机 B 的链路层地址。主机 A 首先发送一个 NS（Neighbor Solicitation，邻居请求）报文，其中源地址为主机 A 的 IPv6 地址，目的地址为主机 B 的被请求节点组播地址，需要解析的目的 IP 地址为主机 A 的 IPv6 地址，表示主机 A 想要知道主机 B 的 MAC 地址。为了便于对方发送应答帧，NS 报文的 Options 字段中携带了主机 A 的 MAC 地址。当主机 B 接收到 NS 报文之后，就会回应 NA（Neighbor Advertisement，邻居公告）报文，其中源地址为主机 B 的 IPv6 地址，目的地址为主机 A 的 IPv6 地址（使用 NS 报文中的主机 A 的链路层地址进行单播），主机 B 的 MAC 地址放在 Options 字段中。

重复地址检测是在接口使用某个 IPv6 单播地址之前进行的，主要是为了探测是否有其他的节点使用了该地址。尤其是在地址自动配置的时候，进行重复地址检测是很必要的。一个 IPv6 单播地址在分配给一个接口之后、通过重复地址检测之前称为试验地址（tentative address）。此时，该接口不能使用这个试验地址进行单播通信，但是仍然会加入两个组播组：All-Nodes 组播组和试验地址所对应的 Solicited-Node 组播组。

IPv6 重复地址检测技术和 IPv4 中的免费 ARP 类似：节点向试验地址所对应的 Solicited-Node 组播组发送 NS 报文。NS 报文中的目的地址即为该试验地址。如果收到其他节点回应的 NA 报文，就证明该地址已在网络上使用，节点将不能使用该试验地址。

RFC4861 中定义了 5 种邻居状态，分别是未完成（INCMP）、可达（REACH）、陈旧（STALE）、延迟（DELAY）、探查（PROBE）。其中，"INCMP" 状态表示邻居不可达，此时正在进行地址解析，邻居的 MAC 地址还未知。若 MAC 解析不成功，则从邻居表中清除；若 MAC 解析成功，则进入 "REACH" 状态。"REACH" 状态表示邻居可达，本机可以直接向处于 "REACH" 状态的邻居发送帧。该邻居若超过老化时间（默认值为 20 分钟）没有再次向本机发送帧，则进入"STALE"状态；本机收到该邻居的 NA 报文且该报文中携带的 MAC 地址发生改变时，也会将该邻居的状态设置为"STALE"。当本机向处于"STALE"状态的邻居发送帧时，需要先发送 NS 报文检测其可达性，一旦发出 NS 报文，便马上将该邻居的状态设置为"DELAY"。默认情况下，"STALE"状态的老化时间是 20 分钟，也就是说，在老化时间内，若本机一直没有收到邻居发送的帧，就将该邻居的状态设置为"DELAY"。当邻居处于"DELAY"状态时，若在老化时间（默认值为 5s）内没收到 NA 报文，则将该邻居的状态设置为"PROBE"。对处于"PROBE"状态的邻居，默认情况下，本机会按照指定的时间间隔（默认值为 1s）向该邻居连续发送 3 个单播 NS 报文，若收到 NA 报文，则将该邻居的状态设置为"REACH"；若无应答则在邻居表中将该邻居删除。

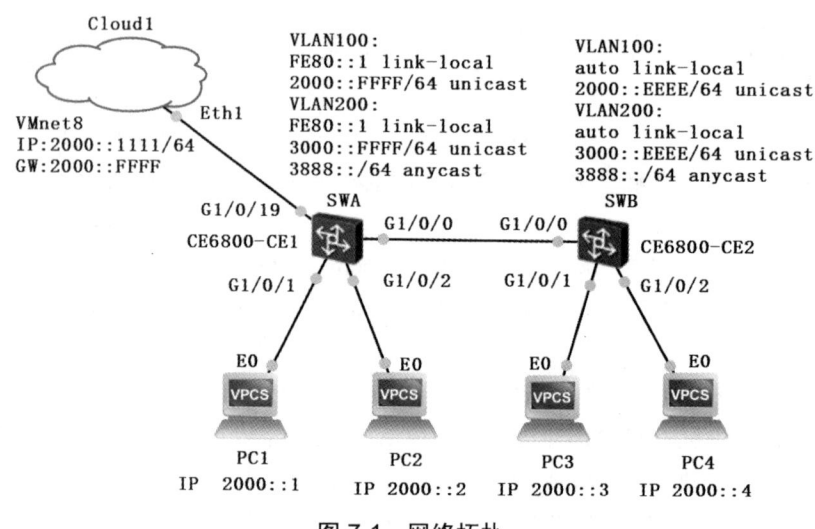

图 7-1 网络拓扑

7.2 仿真实验

7.2.1 仿真环境

仿真环境所需要的硬件、软件建议至少满足以下要求。

- 物理机的处理器主频为 1.6GHz；
- 内存 RAM 大小为 32GB；
- 物理机的操作系统为 Windows 10；
- 虚拟机为 VMware Workstation 16 Pro；
- 网络模拟器选用 GNS3 2.2.31；
- 交换机为华为 CE6800；
- 网络协议分析器为 Wireshark Version 3.4.8；
- PC1~PC4 为 Virtual PC Simulator，Version 0.8.2。

7.2.2 实验设计

根据网络拓扑（图 7-1）完成网络设备的选型：两台交换机选用华为的 CE6800 系列二层交

换机，物理机通过 Cloud1 的 Eth1 接口接入 GNS3 网络中。根据前面的组网需求，本实验设计如下。

1. VLAN 设计

在交换机 SWA（CE6800-CE1）和 SWB（CE6800-CE2）中分别创建两个 VLAN：VLAN100 和 VLAN200，将两台交换机的接口 G1/0/0 设置为 trunk 模式并允许 VLAN100 和 VLAN200 通过。对于交换机 SWA，需要将接口 G1/0/1、G1/0/2 和 G1/0/19 划分到 VLAN100 中，类似地，对于交换机 SWB，需要将接口 G1/0/1、G1/0/2 划分到 VLAN100 中。

2. IP 地址规划

对交换机 SWA 和 SWB 设置两个全局 IP 地址/前缀，分别是 2000::/64 和 3000::/64，设置 1 个任播 IP 地址 3888::/64。各设备接口的 IPv6 地址/前缀如表 7-1 所示，值得注意的是，SWA 的 VLAN100 接口有两个全局 IP 地址。

表 7-1 各设备接口的 IPv6 地址/前缀

设备	接口	全局 IP 地址/前缀	链路本地 IP 地址/前缀	任播 IP 地址/前缀
SWA	VLAN100	2000::FFFF/64 1234::1234/64	FE80::1	—
SWA	VLAN200	3000::FFFF/64	FE80::1	3888::/64
SWB	VLAN100	2000::EEEE/64	auto	—
SWB	VLAN200	3000::EEEE/64	auto	3888::/64
物理机	VMnet8	2000::1111/64	FE80::1	—
PC1	E0	2000::1/64	auto	—
PC2	E0	2000::2/64	auto	—
PC3	E0	2000::3/64	auto	—
PC4	E0	2000::4/64	auto	—

3. 地址解析设计

先清空交换机 SWA 的邻居表，然后在 SWA 上 ping PC2，由于 SWA 的邻居表为空，因此 ping 命令将触发产生用于地址解析的报文对。此时，通过网络协议分析器 Wireshark 捕获到 12 个 ICMPv6 报文，其中前面两个分别是用于地址解析的 NS 报文和 NA 报文。基于这两个报文，可以从 MAC 层、IP 层和 ICMP 协议层面深入剖析 IPv6 基于 ICMPv4 报文的三层地址解析过程。

4. 重复地址检测设计

给交换机 SWA 的 VLAN100 接口设置一个 IPv6 地址 1234::1234/64，此操作将触发产生重复地址检测的报文对。IP 地址 1234::1234 在最终分配给 VLAN100 接口前，将处于试验地址状态，此时交换机将向链路本地内的所有节点发出重复地址检测报文，因为在网络中没有其他接口使用该地址，故检测无冲突，VLAN100 接口成功获得该地址。此时，在网络协议分析器 Wireshark 中，将先后捕获到一对用于重复地址检测的 ICMPv6 报文：NS 报文和 NA 报文。

类似地，给交换机 SWA 的 VLAN100 接口设置一个已经存在的 IPv6 地址"2000::EEEE/64"。很显然，由于 IP 地址 2000::EEEE 已经分配给了 SWB 的接口 VLAN100。因此，必然会检测到冲突，无法通过重复地址检测。由于未能通过重复地址检测，SWA 将不使用该地址，同时将该地址标注为"DUPLICATE"。在网络协议分析器 Wireshark 中，将先后捕获到一对用于重复地址检测的 ICMPv6 报文，即 NS 报文和 NA 报文。基于这两个报文，可以从 MAC 层、IP 层和 ICMP 协议层面，深入剖析 IPv6 基于 ICMPv4 报文的重复地址检测过程。

5. 邻居状态跟踪

邻居表中的"2000::EEEE"的状态依次为"REACH""STALE""DELAY""PROBE""REACH"……，即状态不断循环，若想将此表项超时后从邻居表中删除，则需要关闭 NDP 协议的自动探测功能。

7.2.3 操作步骤

下面给出本次实验的具体操作，一共包括 16 个步骤。

1. 按照图 7-1 所示的网络拓扑，在 GNS3 中新建一个空白的工程

两台交换机选用华为的 CE6800 系列二层交换机，版本为"Version 8.180"，具体型号为"HuaWei CE6800 V200R005C10SPC607B607"；4 台 PC 选用 VPCS。

物理机通过 Cloud 的 Eth1 接口（对应虚拟网卡 VMware Network Adapter VMnet8，IP 地址为 2000::1111/64）接入 GNS3 网络中。

2. 创建 VLAN，并将接口加入 VLAN 中

1）在交换机 SWA 上创建 VLAN100 和 VLAN200，并将接口划分到相应的 VLAN 中

```
<HUAWEI> system-view immediately
[HUAWEI] sysname cxySWA
[cxySWA] vlan batch 100 200
[cxySWA] interface g1/0/0
[cxySWA-GE1/0/0] port link-type trunk
[cxySWA-GE1/0/0] port trunk allow-pass vlan 100 200
[cxySWA-GE1/0/0] quit
[cxySWA]interface g1/0/1
[cxySWA-GE1/0/1] port default vlan 100
[cxySWA]interface g1/0/2
[cxySWA-GE1/0/2] port default vlan 100
[cxySWA]interface g1/0/19
[cxySWA-GE1/0/19] port default vlan 100
```

2）在交换机 SWB 上创建 VLAN100 和 VLAN200，并将接口划分到相应的 VLAN 中

```
<HUAWEI> system-view immediately
[HUAWEI] sysname cxySWB
[cxySWB] vlan batch 100 200
```

[cxySWB] interface g1/0/0

[cxySWB-GE1/0/0] port link-type trunk

[cxySWB-GE1/0/0] port trunk allow-pass vlan 100 200

[cxySWB-GE1/0/0] quit

[cxySWB]interface g1/0/1

[cxySWB-GE1/0/1] port default vlan100

[cxySWB]interface g1/0/2

[cxySWB-GE1/0/2] port default vlan100

3. 在接口上启用 IPv6 功能，并配置接口的链路本地地址

1）手动配置 SWA 的 VLAN100 接口的链路本地地址为 FE80::1

[cxySWA]interface vlan100

[cxySWA-Vlanif100] ipv6 enable

[cxySWA-Vlanif100] ipv6 address FE80::1 link-local

2）手动配置 SWA 的 VLAN200 接口的链路本地地址为 FE80::1

[cxySWA]interface vlan200

[cxySWA-Vlanif200] ipv6 enable

[cxySWA-Vlanif200] ipv6 address FE80::1 link-local

思考：为什么 VLAN100 和 VLAN200 接口的链路本地地址可以设置为相同的？

3）自动配置 SWB 的 VLAN100 接口的链路本地地址

[cxySWB] interface vlan100

[cxySWB-Vlanif100] ipv6 enable

[cxySWB-Vlanif100] ipv6 address auto link-local

4. 查看 SWB 的 VLAN100 接口的链路本地地址

1）查看 SWB 的 VLAN100 接口的 MAC 地址

[cxySWB-Vlanif100] display interface vlan100

Vlanif 100 current state：UP (if index：53)

Line protocol current state：DOWN

Description：

Route Port, The Maximum Transmit Unit is 1500

Internet protocol processing：disabled

IP Sending Frames' Format is PKTFMT_ETHNT_2, **Hardware address is 7021-c3eb-b2cb**

Physical is VLANIF

Current system time：2023-04-0723:22:59

    Last 300 seconds input rate 0 bits/sec， 0 packets/sec

    Last 300 seconds output rate 0bits/sec， 0packets/sec

    Input：0 packets，0 bytes

Output： 0 packets，0 bytes
Last 300 seconds input utility rate：--
Last 300 seconds output utility rate：--

2）查看 SWB 的 VLAN100 接口的链路本地地址

[cxySWB-Vlanif100] display ipv6 interface
Vlanif100 current state：UP
IPv6 protocol current state：UP
IPv6 is enabled, **link-local address is FE80::7221:C3FF:FEEB:B2CB**
　No global unicast_address configured
　Joined group address(es):
　　FF02::1:FFEB:B2CB
　　FF02:: 2
　　FF02::1
MTU is 1500 bytes
ND DAD is enabled，number of DAD attempts： 1
ND NUD is enabled，number of NUD attempts： 3
ND NUD interval is 1000 milliseconds
ND reachable time is 1200000 milliseconds
ND stale time is 1200 seconds
ND retransmit interval is 1000 milliseconds
ND RAs are halted

可见，SWB 的 VLAN100 接口的链路本地地址自动配置为 FE80::7221:C3FF:FEEB:B2CB。

思考：为什么 VLAN100 接口的链路本地地址自动配置为 FE80::7221:C3FF:FEEB:B2CB？当前 VLAN100 接口加入了哪些多播地址？为什么 VLAN100 接口的被请求节点组播地址是 FF02::1:FFEB:B2CB？

5. 从交换机 SWA 上 ping 交换机 SWB 的 VLAN100 接口的链路本地地址

[cxySWA]ping ipv6 FE80::7221:C3FF:FEEB:B2CB
Error: Please specify an interface name for the link-local address

思考：上述命令为什么会出错？即 ping 链路本地地址时为什么需要指定接口名称？

[cxySWA]ping ipv6 FE80::7221:C3FF:FEEB:B2CB -i vlan100
　PING FE80::7221:C3FF:FEEB:B2CB: 56 data bytes，press CTRL_C to break
　　Reply from FE80::7221:C3FF:FEEB:B2CB
　　bytes = 56 Sequence=1hop limit = 64 time = 1249ms
　　Request timeout
　　Reply from FE80::7221:C3FF:FEEB:B2CB
　　bytes = 56 Sequence=3hop limit = 64 time=1957ms

```
    Reply from FE80::7221:C3FF:FEEB:B2CB
    bytes = 56 Sequence=4hop limit = 64 time=1104ms
    Reply from FE80::7221:C3FF:FEEB:B2CB
    bytes = 56 Sequence=5hop limit = 64 time = 894ms
 --- Reply from FE80::7221:C3FF:FEEB:B2CB ping statistics ---
    5 packet(s)  transmitted
    4 packet(s)  received
    20.00% packet loss
    round-trip min/avg/max=894/1301/1957ms
```

可见，交换机 SWA 和 SWB 通过链路本地地址可以实现彼此通信。

6. 配置接口的全球单播地址

1）手动配置 SWA 的 VLAN100 接口的全球单播地址为 2000::FFFF/64

```
[cxySWA] interface vlan100
[cxySWA-Vlanif100] ipv6 address 2000::FFFF/64
```

2）手动配置 SWA 的 VLAN200 接口的全球单播地址为 3000::FFFF/64

```
[cxySWA] interface vlan200
[cxySWA-Vlanif200] ipv6 address 3000::FFFF/64
```

3）手动配置 SWB 的 VLAN100 接口的全球单播地址为 2000::EEEE/64

```
[cxySWB] interface vlan100
[cxySWB-Vlanif100] ipv6 address 2000::EEEE /64
```

4）手动配置 SWB 的 VLAN200 接口的全球单播地址为 3000::EEEE/64

```
[cxySWB]interface vlan200
[cxySWB-Vlanif200] ipv6 address 3000::EEEE/64
```

5）配置 PC1 至 PC4 接口的全球单播地址分别为 2000::1/64 至 2000::4/64

```
PC1> ip 2000::1/64
PC2> ip 2000::2/64
PC3> ip 2000::3/64
PC4> ip 2000::4/64
```

6）配置物理机虚拟网卡 VMware Network Adapter VMnet8 的全球单播地址为 2000::1111/64

```
C:\Windows\system32> netsh interface ipv6 add address interface="21" address=2000::1111/64
```

7）配置物理机虚拟网卡 VMware Network Adapter VMnet8 的默认网关为 2000::FFFF

```
C:\Windows\system32> netsh interface ipv6 add route ::/0  "21" 2000::FFFF
```

上述物理机 VMnet8 接口的 IPv6 地址及网关的设置，也可以通过图形界面完成，此处省略。

7. 查看物理机的 IPv6 路由表

```
C:\Windows\system32> route print -6
接口列表
-------------------------------------------------------------------------------
 10... 98 8f e0 61 18 2b ...... Intel(R) Ethernet Connection (16) I219-V
 22... 02 00 4c 4f 4f 50 ...... Microsoft KM-TEST 环回适配器
 12... e2 0a f6 79 6e e1 ...... Microsoft Wi-Fi Direct Virtual Adapter
 20... f2 0a f6 79 6e e1 ...... Microsoft Wi-Fi Direct Virtual Adapter # 2
 17... 00 50 56 c0 00 01 ...... VMware Virtual Ethernet Adapter for VM net 1
 21... 00 50 56 c0 00 08 ...... VMware Virtual Ethernet Adapter for VM net 8
 18... 00 50 56 c0 00 10 ...... VMware Virtual Ethernet Adapter for VM net 16
  6... e0 0a f6 79 6e e1 ...... Realtek RTL8852BE WiFi 6 802.11ax PCIe Adapter
  1... ........................ Software Loopback Interface1
-------------------------------------------------------------------------------

IPv6 路由表
-------------------------------------------------------------------------------
活动路由:
```

接口	跃点数	网络目标	网关
21	291	::/0	2000::ffff
6	296	::/0	fe80::1
1	331	::1/128	在链路上
21	291	2000::/64	在链路上
21	291	2000::1111/128	在链路上
6	296	2409:8a5e:1419:300::/64	在链路上
6	296	2409:8a5e:1419:300:18bc:4312:c365:566/128	在链路上
6	296	2409:8a5e:1419:300:2943:1ac0:63a4:c891/128	在链路上

8. 从物理机上依次 ping 各个 IPv6 地址

1）从物理机上 ping IP 地址 2000::1

```
C:\Windows\system32>ping 2000::1
正在 Ping 2000::1 具有 32 字节的数据:
请求超时。
来自 2000::1 的回复: 时间=5ms
来自 2000::1 的回复: 时间=2ms
来自 2000::1 的回复: 时间=2ms
2000::1 的 Ping 统计信息:
    数据包: 已发送=4, 已接收=3, 丢失=1 (25%丢失),
往返行程的估计时间(以毫秒为单位):
    最短=2ms, 最长=5ms, 平均=3ms
```

2）类似地，从物理机上 ping 其他 IP 地址：2000::2、2000::FFFF、3000::EEEE

可以发现，命令结果是全部可以 ping 通。

9. 从 PC1 上依次 ping 各个 IPv6 地址

1）从 PC1 上 ping IP 地址 2000::2

```
PC1> ping 2000::2
2000::2 icmp6_seq=1 ttl=64 time=1.961 ms
2000::2 icmp6_seq=2 ttl=64 time=1.694 ms
2000::2 icmp6_seq=3 ttl=64 time=1.213 ms
2000::2 icmp6_seq=4 ttl=64 time=1.971 ms
2000::2 icmp6_seq=5 ttl=64 time=1.014 ms
```

2）从 PC1 上 ping IP 地址：2000::3、2000::1111

可以发现，命令结果是全部可以 ping 通。

3）从 PC1 上 ping IP 地址 3000::FFFF

```
PC1> ping 3000::ffff
host (3000::FFFF) not reachable
```

思考：为什么在物理机上可以 ping 通所有的 IP 地址，在 PC1 上只能 ping 通前缀为 2000::/16 的 IP 地址，而不能 ping 通前缀为 3000::/16 的 IP 地址？

10. 配置接口的任意播地址

1）配置 SWA 的 VLAN200 接口的任意播地址为 3888::/64

```
[cxySWA]interface vlan200
[cxySWA-Vlanif200] ipv6 address 3888::/64 anycast
```

2）配置 SWB 的 VLAN200 接口的任意播地址为 3888::/64

```
[cxySWB]interface vlan200
[cxySWB-Vlanif200]ipv6 address 3888::/64 anycast
```

注意：以上两个接口的 IPv6 地址是相同的，均为 3888::/64，此 IP 地址有何特点？

3）从物理机上 ping 任意播 IPv6 地址 3888::

```
C:\Windows\system32>ping 3888::
正在 Ping3888::具有 32 字节的数据：
来自 3888::的回复：时间=2ms
来自 3888::的回复：时间=1ms
来自 3888::的回复：时间=1ms
来自 3888::的回复：时间=2ms
3888::的 Ping 统计信息：
    数据包：已发送=4，已接收=4，丢失=0 (0%丢失)，
往返行程的估计时间(以毫秒为单位)：
```

最短=1ms，最长=2ms，平均=1ms

思考：此应答报文是由交换机 SWA 还是交换机 SWB 发出的？为什么？

11. 查看 SWA 的 VLAN100 接口的 IPv6 信息

[cxySWA]display ipv6 interface
Vlanif100 current state：UP
IPv6 protocol current state：UP
IPv6 is enabled，link-local address is FE 80::1
　Global unicast address(es)：
　　2000::FFFF，subnet is 2000::/64
　Joined group address(es)：
　　FF02::1:FF00:FFFF
　　FF02::1:FF00:1
　　FF02::2
　　FF02::1
MTU is 1500 bytes
ND DAD is enabled，number of DAD attempts：1
ND NUD is enabled，number of NUD attempts：3
ND NUD interval is 1000 milliseconds
ND reachable time is 1200000 milliseconds
ND stale time is 1200 seconds
ND retransmit interval is 1000 milliseconds
ND RAs are halted

可见，默认情况下，交换机未发送 RA（路由器通告）报文。可在接口视图下执行命令"ipv6 nd ra halt disable"，启用系统发布 RA 报文功能，以实现路由器发现/前缀发现及地址自动配置。现在先不执行该命令。

可在系统视图下执行命令"undo ipv6 nd auto-detect disable"，启用 NDP 协议的自动探测功能，以实现在 PROBE 状态时自动进行不可达检测（NUD）。必要时，可执行命令"ipv6 nd auto-detect disable"禁止 NDP 协议的自动探测功能。默认情况下，该功能是启用的。

12. IPv6 邻居发现协议——地址解析

1）在用户视图下，清除交换机 SWA 的动态邻居表

<cxySWA>reset ipv6 neighbors vlan100
<cxySWA>display ipv6 neighbors　　// 查看交换机 SWA 的邻居表

此时无任何显示信息，说明邻居表为空。

2）在交换机 SWA 中，添加静态邻居表项（PC1 的 IP 地址到 MAC 地址的映射）

[cxySWA-Vlanif100] ipv6 neighbor 2000::1 0050-7966-6801 vlan100 g1/0/1
[cxySWA-Vlanif100] display ipv6 neighbors

```
IPv6 Address        : 2000::1
Link-layer          : 0050-7966-6801        State              : REACH
Interface           : GE1/0/1               Age                : -
VLAN                : 100                   CEVLAN             : -
VPN name            : -                     Is Router          : TRUE
Secure FLAG         : SECURE                Nickname           : -

Total : 1           Dynamic：0              Static：1           Remote：0
```

可见，成功添加了一条静态邻居表项。

3）在交换机 SWA 上 ping PC1

```
[cxySWA-Vlanif100] ping ipv6 2000::1
    PING 2000::1: 56 data bytes，press CTRL_C to break
    Reply from 2000::1
      bytes = 56 Sequence=1 hop limit = 64 time= 1ms
    Reply from 2000::1
      bytes = 56 Sequence=2 hop limit = 64 time = 1ms
    Reply from 2000::1
      bytes = 56 Sequence=3 hop limit = 64 time = 1ms
    Reply from 2000::1
      bytes = 56 Sequence=4 hop limit = 64 time = 1ms
    Reply from 2000::1
      bytes = 56 Sequence=5 hop limit = 64 time = 1ms
```

4）在交换机 SWA 中，添加错误的静态邻居表项

若添加错误的 IP 地址、MAC 地址映射，结果会如何呢？比如将物理机 VMnet8 的 IP 地址由正确的映射 2000::1111→0050-56C0-0008 修改成错误的映射 2000::1111→0050-56C0-4444。

```
[cxySWA-Vlanif100] ipv6 neighbor 2000::1111 0050-56C0-4444 vlan100 g1/0/1
```

5）从物理机上 ping 交换机的 IP 地址 2000::FFFF

```
C:\Windows\system32>ping 2000::FFFF
正在 Ping 2000::ffff 具有 32 字节的数据：
无法访问目标主机。
无法访问目标主机。
无法访问目标主机。
无法访问目标主机。
2000::ffff 的 Ping 统计信息：
    数据包：已发送=4，已接收=0，丢失=4 (100%丢失)
```

可见，此时无法 ping 通交换机，原因是当交换机返回 ICMP 报文时，使用了错误的 MAC 地址封装，致使物理机的 VMnet8 无法收到应答帧。

6）在交换机 SWA 中，删除错误的地址映射，恢复为由动态方式来实现地址解析

[cxySWA-Vlanif100] undo ipv6 neighbor 2000::1111

7）再次从物理机上 ping 交换机的 IP 地址 2000::FFFF

C:\Windows\system32>ping 2000::FFFF
正在 Ping 2000::ffff 具有 32 字节的数据：
来自 2000::ffff 的回复：时间=8ms
来自 2000::ffff 的回复：时间=1ms
来自 2000::ffff 的回复：时间=1ms
来自 2000::ffff 的回复：时间=1ms
2000::ffff 的 Ping 统计信息：
　　数据包：已发送=4，已接收=4，丢失=0 (0%丢失)，
往返行程的估计时间(以毫秒为单位)：
　　最短=1ms，最长=8ms，平均=2ms

8）查看交换机的邻居表

[cxySWA-Vlanif100] display ipv6 neighbors

IPv6 Address : 2000::1111			
Link-layer : 0050-56c0-0008		State : REACH	
Interface : GE1/0/19		Age : 2	
VLAN : 100		CEVLAN : -	
VPN name : -		Is Router : TRUE	
Secure FLAG : UN-SECURE		Nickname : -	
Total：1	**Dynamic：1**	Static：0	Remote：0

可见，交换机 SWA 通过 NDP 协议动态获得了物理机 VMnet8 的正确 MAC 地址解析。

13. 基于 NDP 协议的地址解析

1）捕获用于地址解析的包

（1）清除动态邻居表。

<cxySWA>reset ipv6 neighbors vlan100

（2）准备抓包。在 GNS3 工作区中，右击交换机 SWA（CE6800-CE1）的接口 G1/0/2 和 PC2 之间的链路，然后在弹出菜单中选择"Start capture"命令（见图 7-2），做好抓包的准备。

第 7 章　IPv6 邻居发现协议

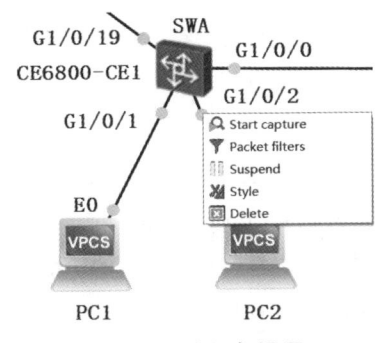

图 7-2　网络抓包设置

（3）在交换机 SWA 上 ping PC2，此操作将触发产生地址解析的报文。

```
<cxySWA>ping ipv6 2000::2
    PING 2000::2：56 data bytes，press CTRL_C to break
    Reply from 2000::2
    bytes = 56 Sequence=1hop limit = 64 time= 264ms
    Reply from 2000::2
    bytes = 56 Sequence=2hop limit = 64 time = 1ms
    Reply from 2000::2
    bytes = 56 Sequence=3hop limit = 64 time = 1ms
    Reply from 2000::2
    bytes = 56 Sequence=4hop limit = 64 time = 1ms
    Reply from 2000::2
    bytes = 56 Sequence=5hop limit = 64 time = 1ms
```

　　ping 命令运行期间，网络协议分析器 Wireshark 将先后捕获到 12 个 ICMPv6 报文，其中前面 2 个分别是用于地址解析的 NS 报文和 NA 报文；后面 10 个是用于连通性测试的"Echo Request"报文和"Echo Reply"报文。

（4）查看交换机 SWA 的邻居表。

```
<cxySWA> display ipv6 neighbors
-------------------------------------------------------------------------
IPv6 Address       : 2000::2
Link-layer         : 0050-7966-6802        State        : REACH
Interface          : GE1/0/2               Age          : 4
VLAN               : 100                   CEVLAN       : -
VPN name           : -                     Is Router    : FALSE
Secure FLAG        : UN-SECURE             Nickname     : -
-------------------------------------------------------------------------
Total : 1          Dynamic : 1             Static : 0   Remote : 0
```

2）根据捕获到的包，分析 NDP 协议的地址解析过程

（1）第一个包——用于地址解析的 NS 报文。用于地址解析的 NS 报文的整体封装层次、帧首部封装、IPv6 首部封装和 ICMPv6 首部封装分别如图 7-3、图 7-4、图 7-5 和图 7-6 所示。请根据图 7-3～图 7-6，分别从 MAC 层、IP 层和 ICMP 协议层面分析 NS 报文的地址解析过程，并回答下列三个问题。

> 请指出源 MAC 地址是什么，并解释说明目的 MAC 地址为什么是 33:33:FF:00:00:02。
> 请指出源 IP 地址是什么，并解释说明目的 IP 地址为什么是 FF02::1:FF00:2。
> 请指出 ICMPv6 的类型是什么，并解释说明目的地址为什么是 2000::2, ICMPv6 选项中的 MAC 地址为什么是 70:51:C0:82:4D:F7。

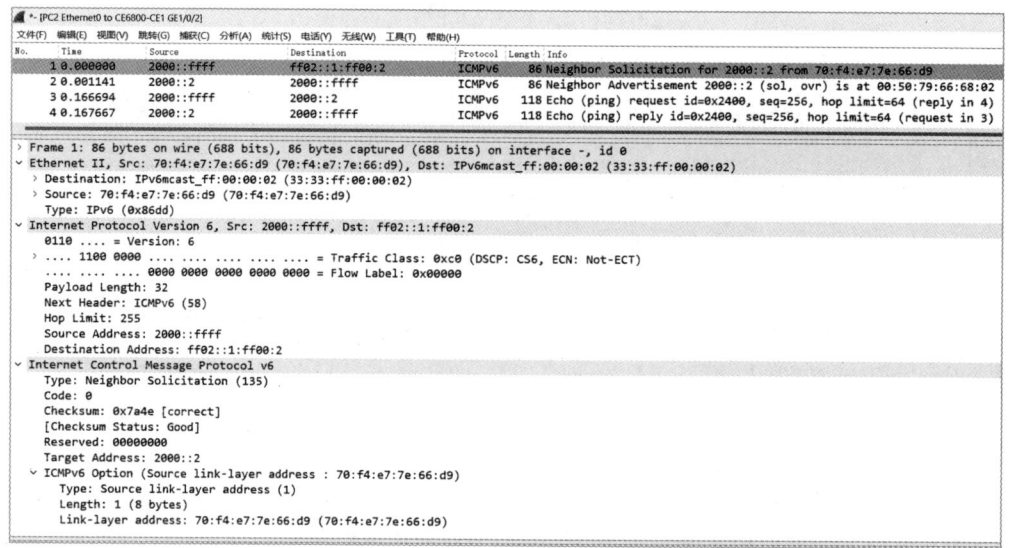

图 7-3　用于地址解析的 NS 报文的整体封装层次

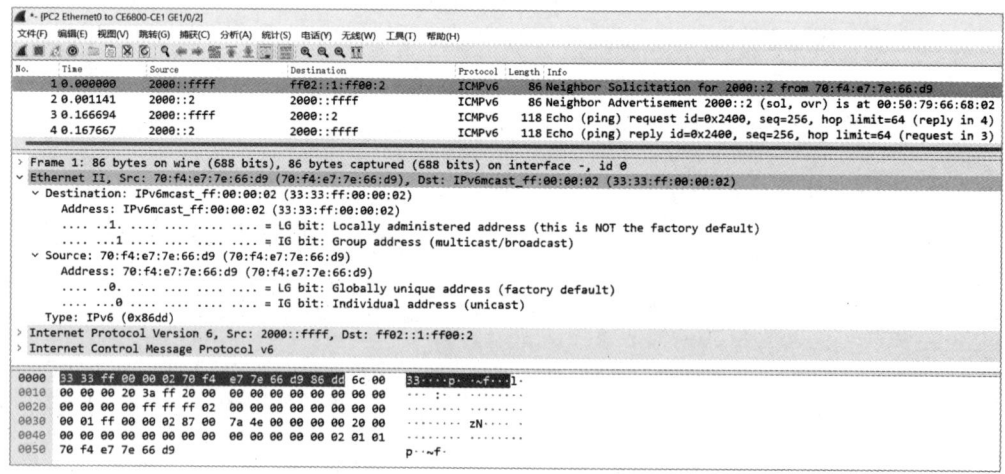

图 7-4　用于地址解析的 NS 报文的帧首部封装

第 7 章　IPv6 邻居发现协议

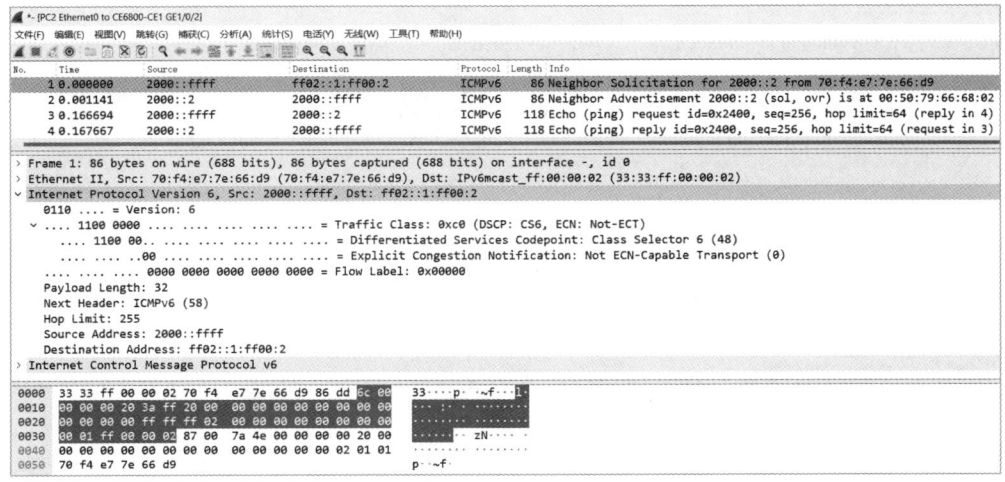

图 7-5　用于地址解析的 NS 报文的 IPv6 首部封装

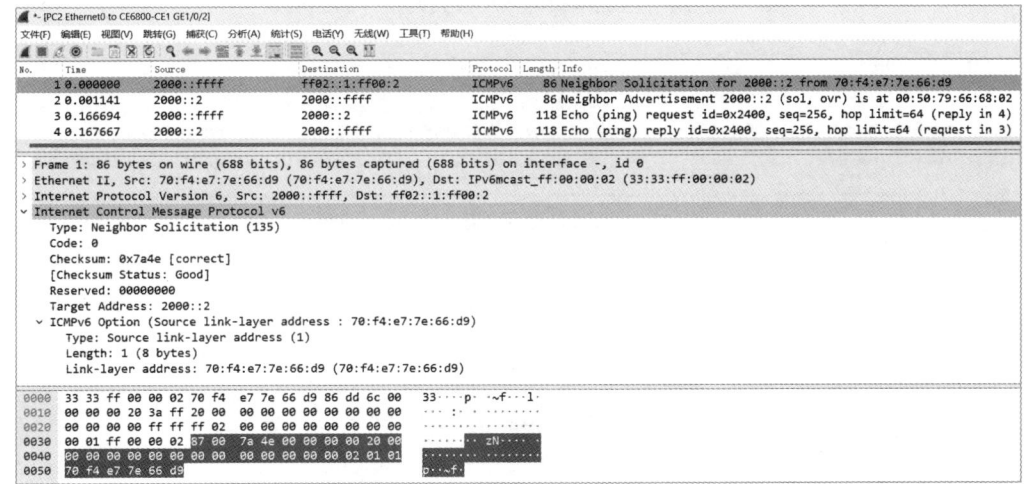

图 7-6　用于地址解析的 NS 报文的 ICMPv6 首部封装

（2）第二个包——用于地址解析的 NA 报文。用于地址解析的 NA 报文的整体封装层次、帧首部封装、IPv6 首部封装和 ICMPv6 首部封装分别如图 7-7、图 7-8、图 7-9 和图 7-10 所示。请根据图 7-7～图 7-10，分别从 MAC 层、IP 层和 ICMP 协议层面分析 NA 报文的地址解析过程，并回答下列三个问题。

> 请指出源 MAC 地址和目的 MAC 地址分别是什么，目的 MAC 地址是单播地址还是组播地址。
> 请指出源 IP 地址和目的 IP 地址分别是什么，目的 IP 地址是单播地址还是组播地址。
> 请指出 ICMPv6 的类型是什么，并解释说明目的地址为什么是 2000::2，ICMPv6 选项中的 MAC 地址为什么是 00:50:79:66:68:02。

至此，IPv6 地址 2000::2 成功解析到 MAC 地址 00:50:79:66:68:02。接下来，交换机 SWA 才能继续完成 ping 过程，即如图 7-10 所示的编号为 3 和 4 的两个 ICMPv6 报文："Echo request" 和 "Echo reply"。

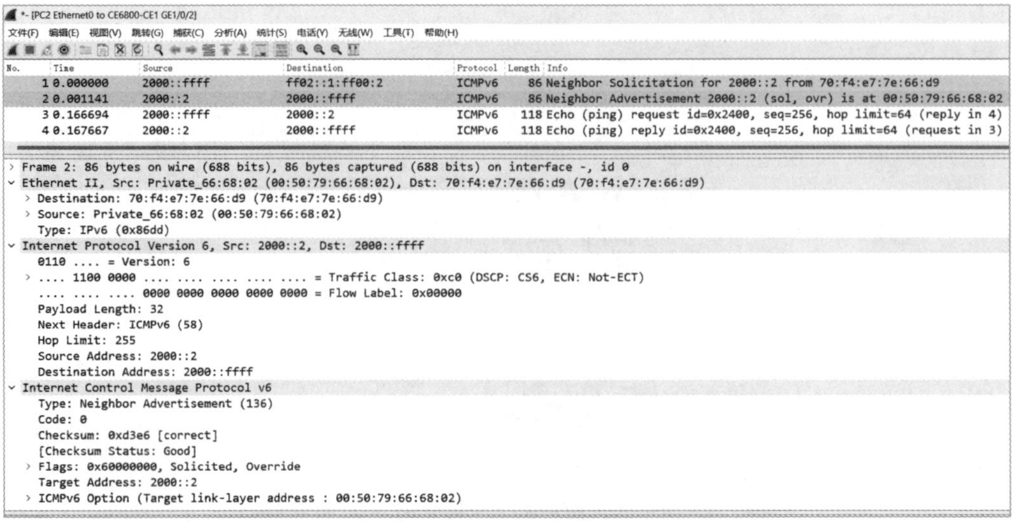

图 7-7 用于地址解析的 NA 报文的整体封装层次

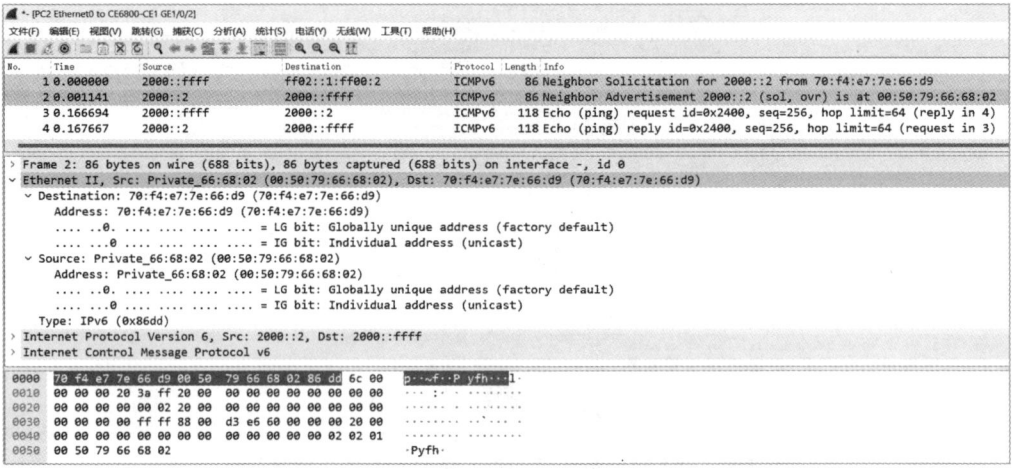

图 7-8 用于地址解析的 NA 报文的帧首部封装

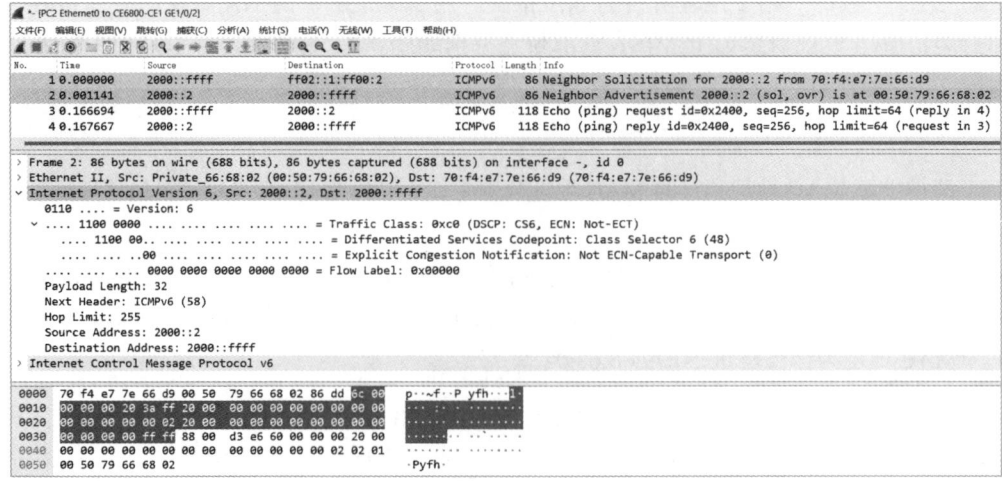

图 7-9 用于地址解析的 NA 报文的 IPv6 首部封装

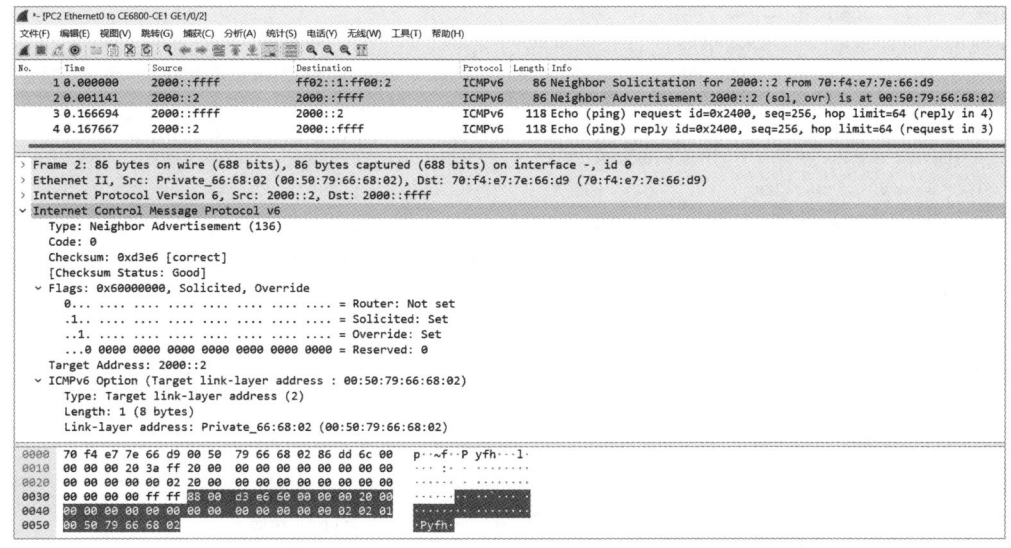

图 7-10　用于地址解析的 NA 报文的 ICMPv6 首部封装

14. 基于 NDP 协议的重复地址检测（地址未冲突，通过重复地址检测）

1）捕获用于重复地址检测的包

（1）准备抓包。在 GNS3 工作区中，右击交换机 SWA 的接口 G1/0/0 和交换机 SWB 的接口 G1/0/0 之间的链路，然后在弹出菜单中选择"Start capture"命令，做好抓包的准备。具体操作界面可参照图 7-2。

（2）为交换机 SWA 的接口 VLAN100 设置另一个 IPv6 地址，此操作不会触发产生重复地址检测的报文对。

> [cxySWA] interface vlan100
> [cxySWA-Vlanif100] ipv6 address 1234::1234/64

IP 地址 1234::1234 在最终分配给 SWA 的接口 VLAN100 前，将处于试验地址状态，此时交换机将向链路本地内的所有节点发出重复地址检测报文，若检测无冲突，则 VLAN100 接口成功获得该地址。网络协议分析器 Wireshark 将先后捕获到一对用于重复地址检测的 ICMPv6 报文，即 NS 报文和 NA 报文。

2）根据捕获到的包，分析 NDP 协议的重复地址检测过程

（1）第一个包——用于重复地址检测的 NS 报文。用于重复地址检测的 NS 报文的整体封装层次、帧首部封装、IPv6 首部封装和 ICMPv6 首部封装分别如图 7-11、图 7-12、图 7-13 和图 7-14 所示。请根据图 7-11～图 7-14，分别从 MAC 层、IP 层和 ICMP 协议层面分析 NS 报文的地址解析过程，并回答下列三个问题。

➢ 请指出源 MAC 地址是什么，并解释说明目的 MAC 地址为什么是 33:33:FF:00:12:34。
➢ 请指出目的 IP 地址是什么，并解释说明源 IP 地址为什么是"::"，目的 IP 地址是单播地址还是组播地址，其值为什么是"FF02::1:FF00:1234"。
➢ 请指出 ICMPv6 的类型是什么，并解释说明目的地址为什么是 1234::1234，有没有 ICMPv6 选项。

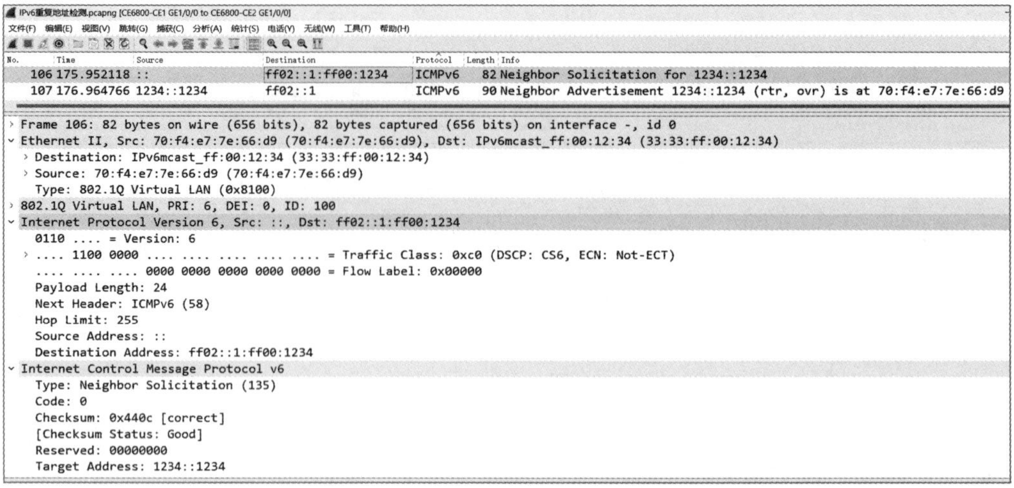

图 7-11 用于重复地址检测的 NS 报文的整体封装层次

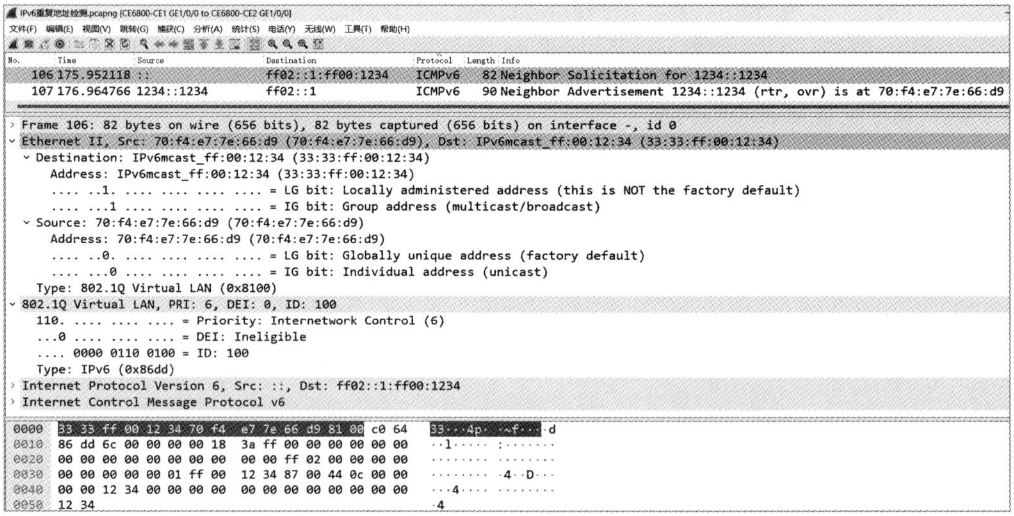

图 7-12 用于重复地址检测的 NS 报文的帧首部封装

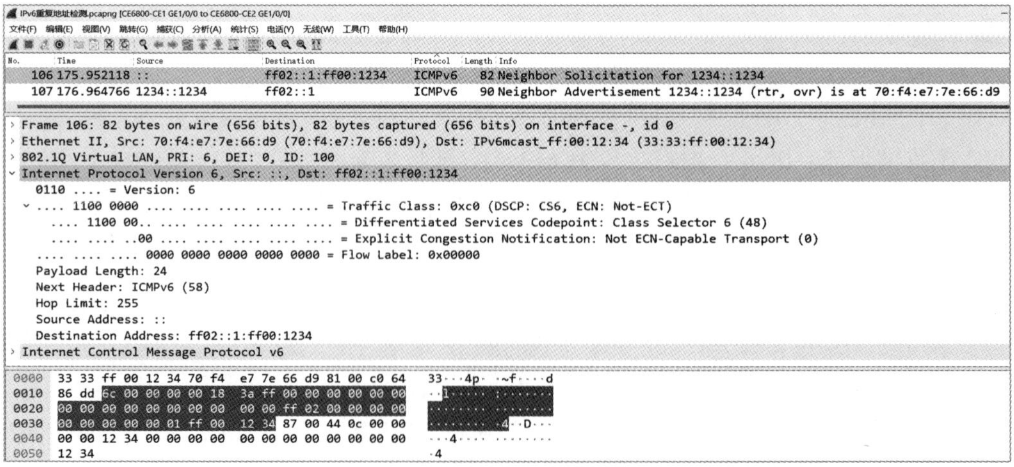

图 7-13 用于重复地址检测的 NS 报文的 IPv6 首部封装

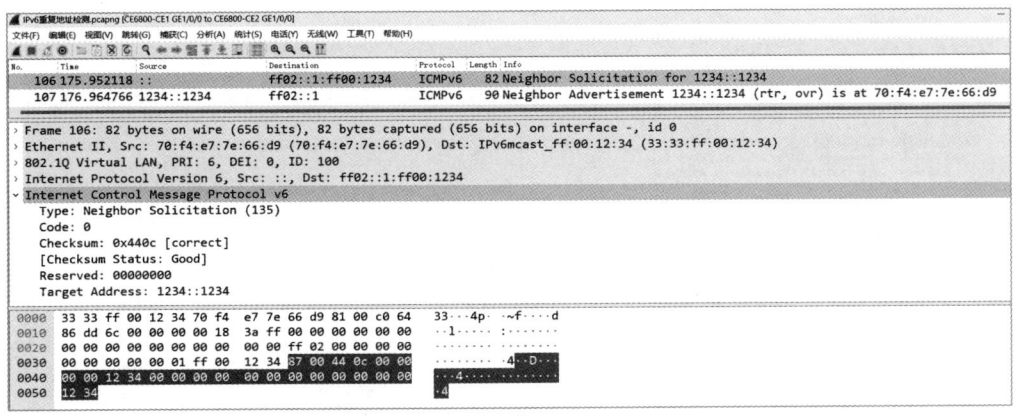

图 7-14　用于重复地址检测的 NS 报文的 ICMPv6 首部封装

（2）第二个包——用于重复地址检测的 NA 报文。若重复地址检测未发现地址冲突，则自己发出 NA 报文，即交换机 VLAN100 接口发出用于重复地址检测的 NA 报文，该报文的整体封装层次、帧首部封装、IPv6 首部封装和 ICMPv6 首部封装分别如图 7-15、图 7-16、图 7-17 和图 7-18 所示。请根据图 7-15～图 7-18，分别从 MAC 层、IP 层和 ICMP 协议层面分析 NA 报文的地址解析过程，并回答下列三个问题。

➢ 请指出源 MAC 地址和目的 MAC 地址分别是什么，目的 MAC 地址是单播地址还是组播地址。

➢ 请指出源 IP 地址和目的 IP 地址分别是什么，目的 IP 地址是单播地址还是组播地址。

➢ 请指出 ICMPv6 的类型是什么，并解释说明目的地址为什么是 2000::2，ICMPv6 选项中的 MAC 地址为什么是 00:50:79:66:68:02。

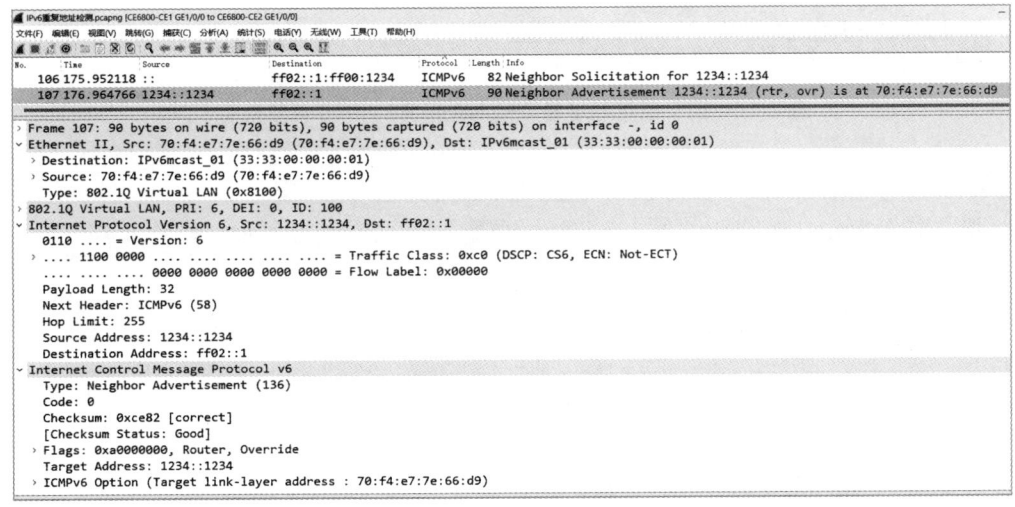

图 7-15　用于重复地址检测的 NA 报文的整体封装层次

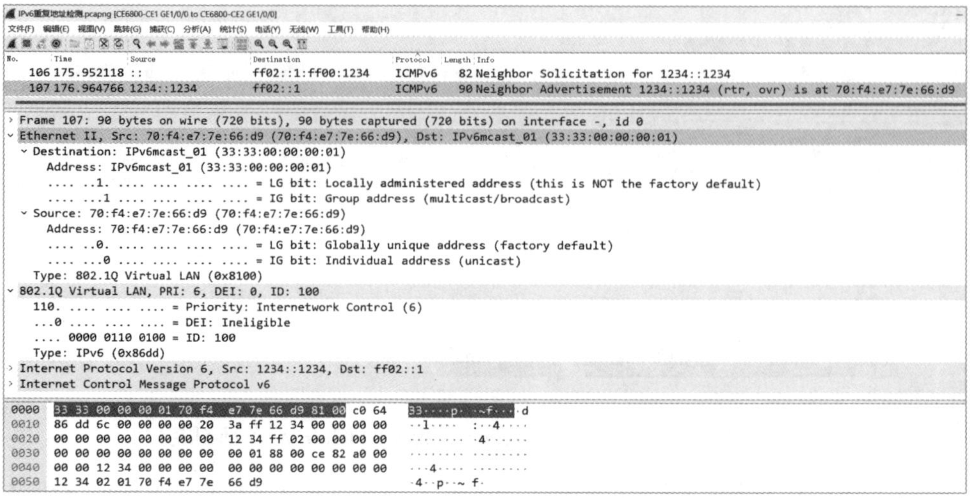

图 7-16 用于重复地址检测的 NA 报文的帧首部封装

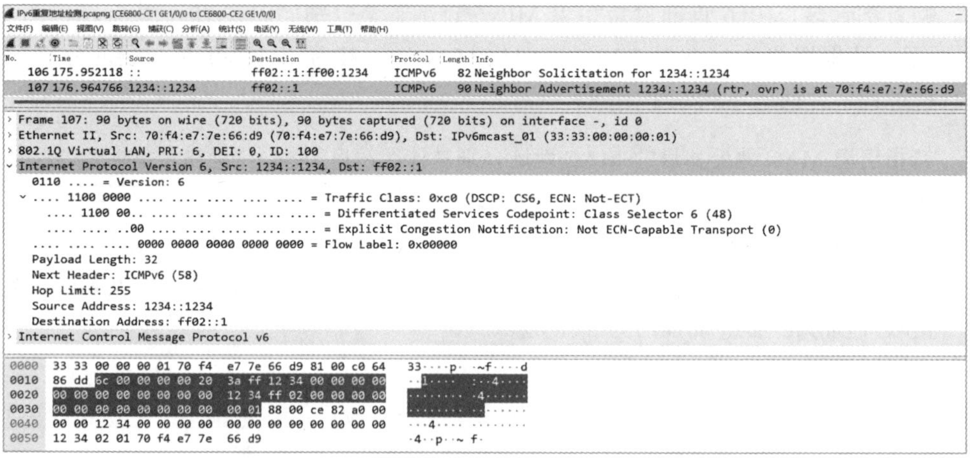

图 7-17 用于重复地址检测的 NA 报文的 IPv6 首部封装

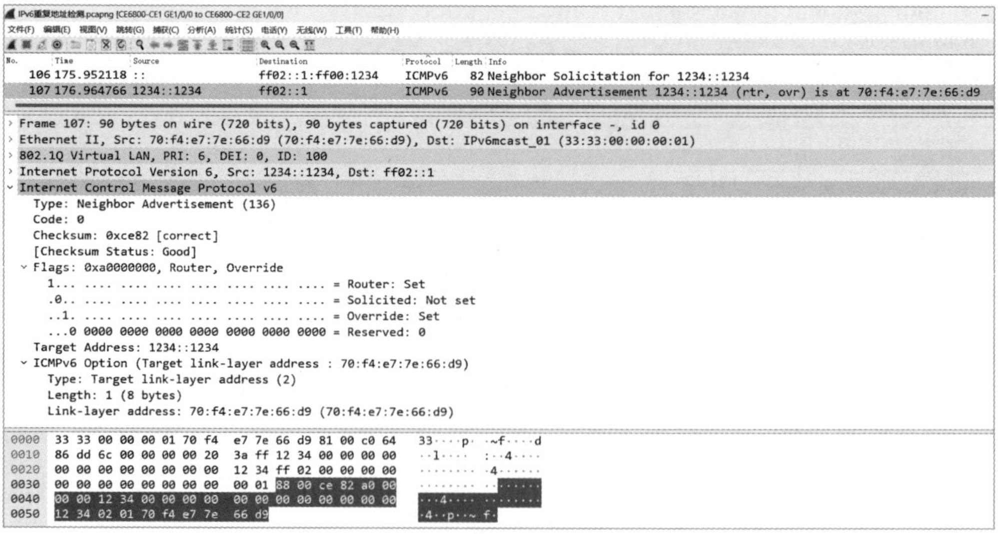

图 7-18 用于重复地址检测的 NA 报文的 ICMPv6 首部封装

15. 基于 NDP 协议的重复地址检测（地址冲突，未通过重复地址检测）

1）捕获用于重复地址检测的包

（1）准备抓包。在 GNS3 工作区中，右击交换机 SWA（CE6800-CE1）的接口 G1/0/0 和交换机 SWB（CE6800-CE2）的接口 G1/0/0 之间的链路，然后在弹出菜单中选择"Start capture"命令，做好抓包的准备。

（2）为交换机 SWA 的接口 VLAN100 设置一个冲突的 IPv6 地址，此操作将触发产生重复地址检测的报文对。

> [cxySWA] interface vlan100
> [cxySWA-Vlanif100] undo ipv6 address 2000::FFFF/64
> // 设置接口的 IPv6 地址为 2000::EEEE/64，注意，该地址与 SWB 的 VLAN100 接口的地址相同
> [cxySWA-Vlanif100] ipv6 address 2000::EEEE/64

IP 地址 2000::EEEE 在最终分配给 SWA 的接口 VLAN100 前，处于试验地址状态，此时交换机将向链路本地内的所有节点发出重复地址检测报文。很显然，IP 地址 2000::EEEE 已经分配给了 SWB 的 VLAN100 接口。因此，必然会检测到冲突，无法通过重复地址检测。网络协议分析器 Wireshark 将先后捕获到一对用于重复地址检测的 ICMPv6 报文，即 NS 报文和 NA 报文。

2）根据捕获到的包，分析 NDP 协议的重复地址检测过程

（1）第一个包——用于重复地址检测的 NS 报文。用于重复地址检测的 NS 报文的整体封装层次如图 7-19 所示。请根据图 7-19，分别从 MAC 层、IP 层和 ICMP 协议层面分析 NS 报文的重复地址检测过程，并回答下列问题。

➤ 请指出源 MAC 地址是什么，并解释说明目的 MAC 地址为什么是 33:33:FF:00:EE:EE。
➤ 请指出目的 IP 地址是什么，并解释说明源 IP 地址为什么是"::"；目的 IP 地址是单播地址还是组播地址，其值为什么是"FF02::1:FF00:EEEE"。
➤ 请指出 ICMPv6 的类型是什么，并解释说明目的地址为什么是 2000::EEEE。

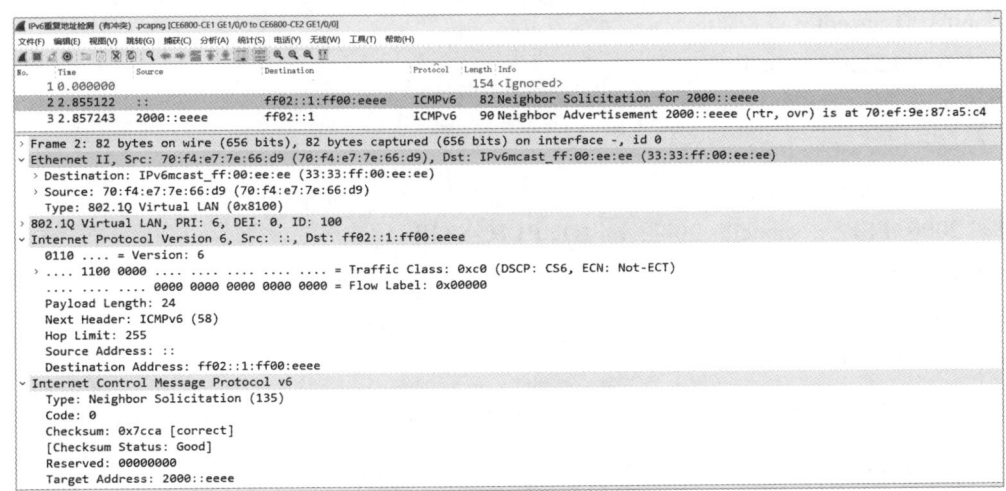

图 7-19　用于重复地址检测的 NS 报文的整体封装层次（地址冲突）

（2）第二个包——用于重复地址检测的 NA 报文。重复地址检测发现接口地址与 SWB 的 VLAN100 接口地址冲突，此时由 SWB 发出 NA 报文，该报文的整体封装层次如图 7-20 所示。SWA 收到此报文后，将不使用 IP 地址 2000::EEEE 并标注该地址为 "DUPLICATE"。请根据图 7-20，分别从 MAC 层、IP 层和 ICMP 协议层面分析 NA 报文的重复地址检测过程，并回答下列问题。

> 请指出源 MAC 地址是什么，并解释说明目的 MAC 地址为什么是 33:33:00:00:00:01。
> 请解释说明源 IP 地址为 2000::EEEE 意味着什么，目的 IP 地址为什么是 FF02::1。
> 请指出 ICMPv6 的类型是什么，并解释说明目的地址为什么是 2000::EEEE，ICMPv6 选项中的 MAC 地址为什么是 70:EF:9E:87:A5:C4（该地址是哪个接口的）。

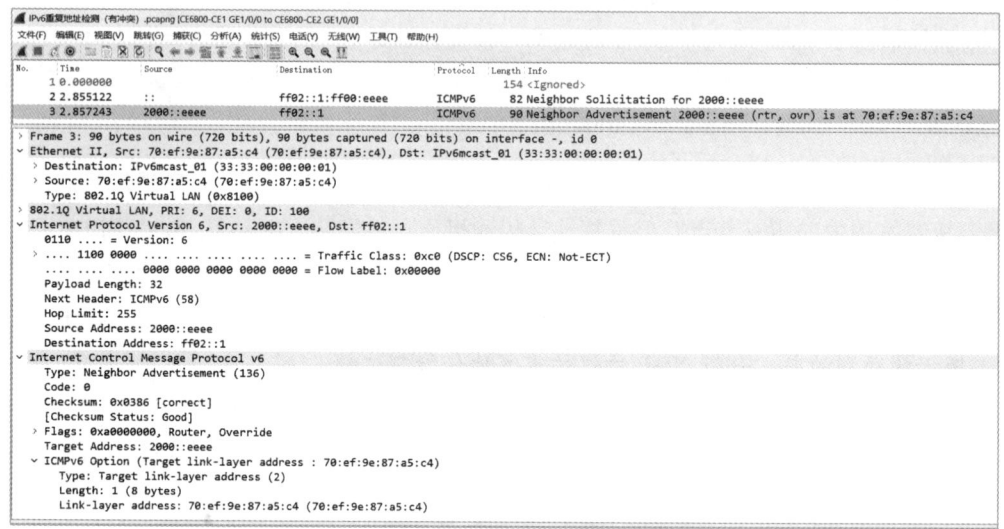

图 7-20　用于重复地址检测的 NA 报文的整体封装层次（地址冲突）

（3）在交换机 SWA 上查看 VLAN100 接口的 IP 地址。

[cxySWA] display ipv6 interface vlan100

Vlanif 100 current state：UP

IPv6 protocol current state：UP

IPv6 is enabled，link-local address is FE80::1

　Global unicast address(es)：

　　1234::1234，subnet is 1234::/64

　　2000::EEEE，subnet is 2000::/64 **[DUPLICATE]**

　Joined group address(es)：

　　FF02::1: FF00: EEEE

　　FF02::1: FF00:1234

　　FF02::1: FF00:1

　　FF02::2

　　FF02::1

　　……

可见，由于 IP 地址存在冲突，故交换机 SWA 不使用 IP 地址 2000::EEEE，并标注该地址为"DUPLICATE"。

16. IPv6 邻居发现协议——邻居状态跟踪

1）INCMP 状态

（1）设置交换机 SWA 的 VLAN100 接口的邻居可达性探测的时间间隔为 20s。

[cxySWA-Vlanif100] ipv6 nd ns retrans-timer 20000

（2）在 SWA 和 SWB 链路上设置抓包。

（3）在交换机 SWA 上 ping 一个不存在的 IPv6 地址 2000::8888。

[cxySWA-Vlanif100] ping ipv6 2000::8888　　#显然不会通

（4）在交换机 SWA 上查看邻居表。

```
[cxySWA-Vlanif100] display ipv6 neighbors
------------------------------------------------------------------------
IPv6 Address      : 2000::8888
Link-layer        : 0000-0000-0000       State        : INCMP
Interface         : Vlanif100            Age          : 0
VLAN              : -                    CEVLAN       : -
VPN name          : -                    Is Router    : FALSE
Secure FLAG       : UN-SECURE            Nickname     : -
------------------------------------------------------------------------
Total: 1   Dynamic: 1    Static: 0    Remote: 0
```

可见，此时 SWA 在邻居表中新建了一个表项 2000::8888，状态为"INCMP"。经过片刻，再显示邻居表，就会发现没有该表项了，即将该表项从本机的邻居表中删除。

（5）抓包分析。

如图 7-21 所示，SWA 每隔 20s 向 2000::8888 陆续发出 3 个 NS 报文。其中，时间间隔 20s 是由命令"ipv6 nd ns retrans-timer 20000"设置的。

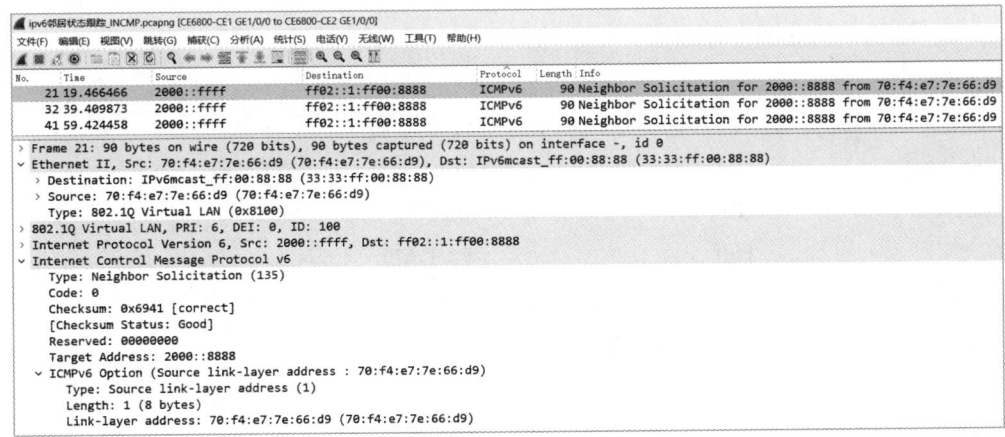

图 7-21　NS 报文的发送时间间隔

2) REACH 状态

(1) 设置 REACH 状态的老化时间为 3min（单位为 ms）。

[cxySWA-Vlanif100] ipv6 nd nud reachable-time 180000

(2) 清空邻居表。

\<cxySWA\> reset ipv6 neighbors vlan100

(3) 在交换机 SWA 上 ping IPv6 地址 2000::EEEE。

\<cxySWA\>ping ipv6 2000::EEEE

(4) 在交换机 SWA 上查看邻居表。

\<cxySWA\>display ipv6 neighbors

IPv6 Address	: 2000::EEEE		
Link-layer	: 70ef-9e87-a5c4	State	: REACH
Interface	: GE1/0/0	Age	: 0
VLAN	: 100	CEVLAN	: -
VPN name	: -	Is Router	: TRUE
Secure FLAG	: UN-SECURE	Nickname	: -

Total: 1   Dynamic: 1   Static: 0   Remote: 0

3) STALE 状态

(1) 设置 STALE 的老化时间为 3min（单位为 s）。

[cxySWA-Vlanif100] ipv6 nd stale-timeout 180

(2) 在交换机 SWA 上查看邻居表。

\<cxySWA\>display ipv6 neighbors

IPv6 Address	: 2000::EEEE		
Link-layer	: 70ef-9e87-a5c4	State	: STALE
Interface	: GE1/0/0	Age	: 1
VLAN	: 100	CEVLAN	: -
VPN name	: -	Is Router	: TRUE
Secure FLAG	: UN-SECURE	Nickname	: -

Total: 1   Dynamic: 1   Static: 0   Remote: 0

4）DELAY 状态

在交换机 SWA 上查看邻居表。

```
<cxySWA>display ipv6 neighbors
------------------------------------------------------------------------
IPv6 Address      : 2000::EEEE
Link-layer        : 70ef-9e87-a5c4      State       : DELAY
Interface         : GE1/0/0             Age         : 6
VLAN              : 100                 CEVLAN      : -
VPN name          : -                   Is Router   : TRUE
Secure FLAG       : UN-SECURE           Nickname    : :-
------------------------------------------------------------------------
Total: 1    Dynamic: 1    Static: 0    Remote: 0
```

5）PROBE 状态

（1）设置邻居表 PROBE 状态的探测次数为 5 次。

```
[cxySWA-Vlanif100] ipv6 nd nud attempts 5
```

（2）设置邻居表 PROBE 状态的时间间隔为 10s（单位为 ms）。

```
[cxySWA-Vlanif100] ipv6 nd nud interval 10000
```

（3）在交换机 SWA 上查看邻居表。

```
<cxySWA> display ipv6 neighbors
------------------------------------------------------------------------
IPv6 Address      : 2000::EEEE
Link-layer        : 70ef-9e87-a5c4      State       : PROBE
Interface         : GE1/0/0             Age         : 6
VLAN              : 100                 CEVLAN      : -
VPN name          : -                   Is Router   : TRUE
Secure FLAG       : UN-SECURE           Nickname    : :-
------------------------------------------------------------------------
Total: 1    Dynamic: 1    Static: 0    Remote: 0
```

可见，邻居表中的 2000::EEEE 的状态依次经过了"REACH""STALE""DELAY""PROBE""REACH"……即状态不断循环，若想将此表项超时后从邻居表中删除，则需要关闭 NDP 协议的自动探测功能。可以在系统视图中运行命令"ipv6 nd auto-detect disable"来关闭 NDP 协议的自动探测功能（该功能默认为启用状态）。

第3篇 路由器安全篇

第 8 章　IPv6 无状态地址自动配置

第 9 章　DHCPv6——地址自动分配

第 10 章　基于路由器的 CBAC 防火墙

第 11 章　基于路由器的 IPsec VPN

第 12 章　基于路由器的 Easy VPN

第 8 章
IPv6 无状态地址自动配置

8.1 组网需求

网络拓扑如图 8-1 所示，要求主机 PC1 通过 SLAAC（StateLess Address AutoConfiguration，无状态地址自动配置）方式从路由器 R1 处获得全球 IPv6 地址和默认网关。在链路范围内，路由器 R1 和 R2 均开启了 RA 功能，现要求路由器 R3 的 G0/0/1 接口和物理机 VMnet8 接口都通过 SLAAC 方式从 R1 或 R2 处获得全球 IPv6 地址和默认网关。当 R1 和 R2 优先级相同时，主机的默认网关设置成二者；当优先级不同时，则选择优先级高的路由器作为自己的网关。

IPv6 主机无状态自动配置主要包括以下几个过程：根据接口标识产生链路本地地址，发出邻居请求，进行重复地址检测，若通过重复地址检测，则链路本地地址生效，至此节点具备本地链路通信能力。此时，主机会向本链路内的所有路由器发送 RS 报文，然后主机会接收到 RA 报文。根据 RA 报文中携带的前缀信息和本机接口标识即可自动生成一个全局 IPv6 地址，并获得其他相关信息。

在 RA 报文中有管理地址配置标志位 MFLAG 和其他状态配置标志位 OFLAG。当 RA 报文中的 M=0 且 O=0 时，路由器告诉主机要通过 SLAAC 方式获取前缀信息，并获取当前跳数限制、路由器生存时间、可达时间和重传定时器等其他网络参数的值。

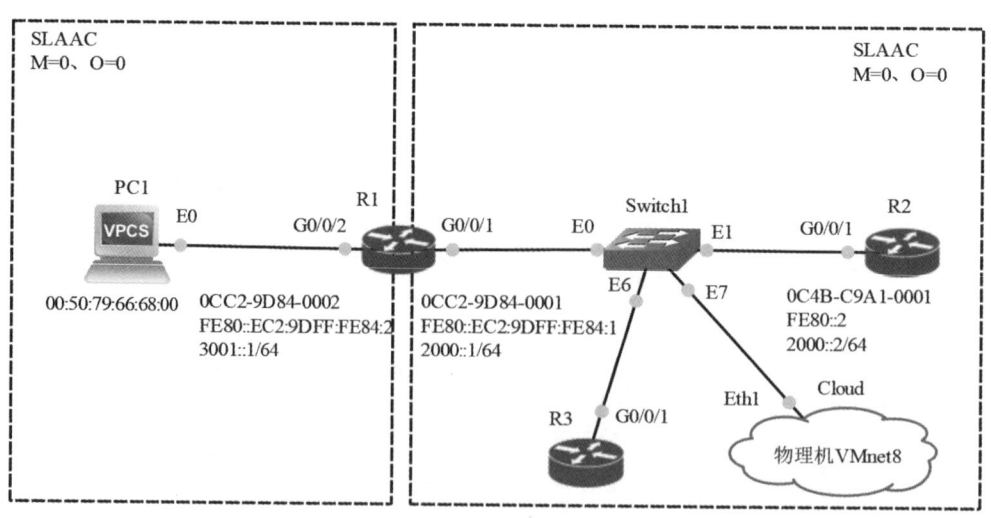

图 8-1　网络拓扑

8.2 仿真实验

8.2.1 仿真环境

仿真环境所需要的硬件、软件建议至少满足以下要求。

- 物理机的处理器主频为 1.6GHz；
- 内存 RAM 大小为 32.0GB；
- 物理机的操作系统为 Windows 10；
- 虚拟机为 VMware Workstation 16 Pro；
- 网络模拟器选用 GNS3 2.2.31；
- 交换机为 Ethernet Switch；
- 路由器为 Huawei AR1kV；
- PC1 为 Virtual PC Simulator, Version 0.8.2。

8.2.2 实验设计

根据网络拓扑（见图 8-1）完成网络设备的选型：3 台路由器均为华为 AR 系列路由器，物理机通过 Cloud 的 Eth1 接口接入 GNS3 网络中。根据前面的组网需求，本实验设计如下。

1. VLAN 设计

交换机无须创建新的 VLAN，所有接口均属于 VLAN1。

2. IP 地址规划

网络中有两个全局 IP 地址/前缀，分别是 2000::/64 和 3000::/64，各设备接口的 IPv6 地址/前缀如表 8-1 所示。

表 8-1 各设备接口的 IPv6 地址/前缀

设备	接口	全局 IP 地址/前缀	链路本地 IP 地址/前缀	任播 IP 地址/前缀
R1	G0/0/1	2000::1/64	auto	—
R1	G0/0/2	3001::1/64	auto	—
R2	G0/0/1	2000::2/64	FE80::2	—
R3	G0/0/1	IP 地址通过 SLAAC 方式获得	auto	—
物理机	VMnet8	IP 地址通过 SLAAC 方式获得	auto	—
PC1	E0	IP 地址通过 SLAAC 方式获得	auto	—

3. 修改 RA 报文字段默认值

启用路由器 R1 接口 G0/0/2 发送 RA 报文功能，并修改 RA 报文中的重传定时器、当前跳数限制、可达时间字段的值。在此基础上，修改路由器生存时间为 900s。默认情况下，路由器生存时间为 RA 报文的最大发布间隔时间的 3 倍（1800s）。

在链路本地范围内有两台路由器启用 RA，物理机的 VMnet8 接口和 R3 路由器的接口通过 SLAAC 的方式获取前缀信息，生成全局 IPv6 地址。若这两个路由器的默认优先级是相同的，则待分配 IP 地址的客户机将获得两个默认网关；若这两个路由器的优先级不同，则客户机选择优先级高的路由器作为自己的网关。

8.2.3 操作步骤

下面给出本次实验的具体操作步骤。

1. 按照图 8-1 所示的网络拓扑，在 GNS3 中新建一个空白的工程

3 台路由器 R1 至 R3 均为华为系列路由器，具体型号为"Huawei AR1kV V300R021C00SPC100T"，一台 PC 选用 VPCS。

物理机通过 Cloud 的 Eth1 接口（对应虚拟网卡 VMware Network Adapter VMnet8，IP 地址为 192.168.241.1/24）接入 GNS3 网络中。

2. 路由器首次启动需要完成用户名及密码的配置

1）启动路由器（启动完成后出现以下信息）

```
netconf monitor start!
    Vrp Bsp so creation Data：Aug 12 2021，15:07:59
    FLASH_Create Flash File /mnt/misc/flash file
DOPRA initialize...
Info：The current cpu freq is 2687997952
OK
IPSI CRYPTO selftest..........................OK
IAS initialize................................OK
VRP initialize
Connect to cap................................OK
Create tasks..................................OK
Initialize tasks..............................
Reade label result............................0x810401ab
Starting the server...
OK
Recovering configuration..................... OK
    Press any key to get started
```

2）按任意键，然后根据提示设置登录设备的用户名和密码（至少 8 位，且含字母、数字或特殊符号）

```
    Press any key to get started
    Login authentication
    Warning：An initial username and password are required for the first login via the console.
    Set a username and password and keep them safe. Otherwise you will not be able to login via the console.
    New Username：cxy
    Password：
    Confirm password：
```

The account create success.
Info：Configuration console exit，please retry to log on

3）重新登录，并输入刚才设置的用户名和密码

Login authentication
Username：cxy
Password：
<Huawei>

4）在出现警告信息后按"Y"键，停止自动配置，进入用户视图

Warning：Auto-Config is working. Before configuring the device，stop Auto-Config. If you perform configurations when Auto-Config is running，the DHCP, routing, DNS, and VTY configurations will be lost. Do you want to stop Auto-Config? [y/n]：
Apr 21 2023 04:07:06+00:00 Huawei LINE/4/USERLOGIN: OID1.3.6.1.4.1.2011.5.25.207.2.2 A use rlogin. (User Index=0，UserName=cxy，User IP=Console0，UserChannel=CON0)
<Huawei>y
Info：Auto-Config has been stopped.
<Huawei>

3. 链路本地范围内只有一台路由器启用发送 RA 报文功能，PC1 通过 SLAAC 方式获取前缀信息，生成全局 IPv6 地址

1）将路由器 R1 命名为 cxyR1（可将 cxy 改为你的名字拼音首字母）

<Huawei>system-view
Enter system view, return user view with Ctrl+Z.
[Huawei] sysname cxyR1
[cxyR1]

2）配置从 console 登录设备时永不超时

[cxyR1]user-interface console 0
[cxyR2-ui-console0] idle-timeout 0
Info: The timeout period is 0, the terminal will remain in the login state, leading to security risks.

3）设置接口 IPv6 地址

[cxyR1] ipv6   // 在系统视图启用 IPv6
[cxyR1] interface g0/0/2
[cxyR1-GigabitEthernet0/0/2] ipv6 enable   // 在接口视图启用 IPv6
[cxyR1-GigabitEthernet0/0/2] ipv6 address 3001::1/64

4）开启路由器 R1 接口 G0/0/2 的发送 RA 报文功能

此功能默认是关闭的，因此若在本链路范围内没有路由器启用发送 RA 报文功能，则当主机设置为自动获得全局地址时将会出错。例如，在 PC1 上运行"ip auto"时，将出现"No router answered ICMPv6 Router Solicitation"的提示信息。

[cxyR1-GigabitEthernet0/0/2] undo ipv6 nd ra halt

5）查看 R1 接口 G0/0/2 的信息

[cxyR1-GigabitEthernet0/0/2] display ipv6 interface g0/0/2
GigabitEthernet 0/0/ 2 current state：UP
IPv6 protocol current state：UP
IPv6 is enabled，link-local address is FE 80::EC2:9DFF:FE84:2
　Global unicast address(es)：
　　3001::1，subnet is 3001::/ 64
　Joined group address(es)：
　　FF02：：1：FF00：1
　　FF02：：2
　　FF02：：1
　　FF02：：1：FF84：2
MTU is 1500 bytes
ND DAD is enabled，number of DAD attempts：1
ND reachable time is 30000 milliseconds
ND retransmit interval is 1000 milliseconds
ND stale time is 1200 seconds
ND advertised reachable time is 0 milliseconds
ND advertised retransmit interval is 0 milliseconds
ND router advertisement max interval 600 seconds，min interval 200 seconds
ND router advertisements live for 1800 seconds
ND router advertisement shop-limit 64
ND default router preference medium
Hosts use stateless autoconfig for addresses

6）在 PC1 和 R1 之间的链路上设置抓包，设置 PC1 自动获得全局 IP 地址

PC1> ip auto
GLOBAL SCOPE　　　　　：3001::2050:79FF:FE66:6800/64
ROUTER LINK-LAYER　　：0c:c2:9d:84:00:02

可见，运行命令"undo ipv6 nd ra halt"启用发送 RA 报文功能后，路由器即可通过 SLAAC 方式向主机分配 IP 地址。此时，PC1 通过无状态方式分配得到了全局 IPv6 地址：3001::2050:79FF:FE66:6800。

7）对通过 SLAAC 方式获得的 IP 地址进行报文分析

（1）RS 报文。

用 SLAAC 方式自动分配 IP 地址的 RS 报文的结构如图 8-2 所示，请分别从 MAC 层、IP 层、ICMP 协议封装的角度对 RS 报文进行分析，并回答下列问题。

➢ 源 MAC 地址、目的 MAC 地址是单播地址还是组播地址，是本地的还是全球的？
➢ IP 分组的封装格式（包括版本、通信量类、流标记、载荷长度、下一首部、跳数限制、源 IP 地址、目的 IP 地址）是什么？
➢ 目的 IP 地址是如何映射到目的 MAC 地址的？
➢ ICMPv6 报文的类型、代码、校验和分别是什么？

```
文件(F) 编辑(E) 视图(V) 跳转(G) 捕获(C) 分析(A) 统计(S) 电话(Y) 无线(W) 工具(T) 帮助(H)
No.     Time        Source              Destination        Protocol   Length Info
  1 0.000000        3001::2050:79ff:fe6… ff02::2           ICMPv6     62 Router Solicitation
  2 0.838197        fe80::ec2:9dff:fe84… ff02::1           ICMPv6     142 Router Advertisement from 0c:c2:9d:84:00:02

> Frame 1: 62 bytes on wire (496 bits), 62 bytes captured (496 bits) on interface -, id 0
∨ Ethernet II, Src: Private_66:68:00 (00:50:79:66:68:00), Dst: IPv6mcast_02 (33:33:00:00:00:02)
  ∨ Destination: IPv6mcast_02 (33:33:00:00:00:02)
      Address: IPv6mcast_02 (33:33:00:00:00:02)
      .... ..1. .... .... .... .... = LG bit: Locally administered address (this is NOT the factory default)
      .... ...1 .... .... .... .... = IG bit: Group address (multicast/broadcast)
  ∨ Source: Private_66:68:00 (00:50:79:66:68:00)
      Address: Private_66:68:00 (00:50:79:66:68:00)
      .... ..0. .... .... .... .... = LG bit: Globally unique address (factory default)
      .... ...0 .... .... .... .... = IG bit: Individual address (unicast)
    Type: IPv6 (0x86dd)
∨ Internet Protocol Version 6, Src: 3001::2050:79ff:fe66:6800, Dst: ff02::2
    0110 .... = Version: 6
  ∨ .... 0000 0000 .... .... .... .... .... = Traffic Class: 0x00 (DSCP: CS0, ECN: Not-ECT)
      .... 0000 00.. .... .... .... .... .... = Differentiated Services Codepoint: Default (0)
      .... .... ..00 .... .... .... .... .... = Explicit Congestion Notification: Not ECN-Capable Transport (0)
    .... .... .... 0000 0000 0000 0000 0000 = Flow Label: 0x00000
    Payload Length: 8
    Next Header: ICMPv6 (58)
    Hop Limit: 255
    Source Address: 3001::2050:79ff:fe66:6800
    Destination Address: ff02::2
∨ Internet Control Message Protocol v6
    Type: Router Solicitation (133)
    Code: 0
    Checksum: 0x4b00 [correct]
    [Checksum Status: Good]
    Reserved: 00000000
```

图 8-2 RS 报文的结构

（2）RA 报文。

用 SLAAC 方式自动分配 IP 地址的 RA 报文的结构如图 8-3 所示，RA 报文的 MAC 层和 IP 层封装如图 8-4 所示，RA 报文的 ICMPv6 封装如图 8-5 和图 8-6 所示。请根据图 8-3～图 8-6，分别从 MAC 层、IP 层、ICMP 协议封装的角度对 RA 报文进行分析，并回答下列问题。

➢ 源 MAC 地址、目的 MAC 地址是单播地址还是组播地址，是本地的还是全球的？
➢ IP 分组的源 IP 地址、目的 IP 地址分别是什么？目的 IP 地址是如何映射到目的 MAC 地址的？
➢ ICMPv6 报文的类型、代码、校验和分别是什么？
➢ ICMPv6 报文中的"Cur hop limit"的值是多少？代表什么意思？
➢ ICMPv6 报文中的 M、O 位的取值分别是什么？代表什么意思？
➢ ICMPv6 报文中的"Router lifetime""Reachable time""Retrans timer"的值各是多少？代表什么意思？
➢ ICMPv6 报文中有几个 ICMPv6 选项？分别是什么？
➢ 前缀信息"Prefix information"选项中的 L、A、R 位的取值分别是什么？代表什么意思？有效生存时间"Valid Lifetime"和首选生存时间"Preferred Lifetime"分别为多长？

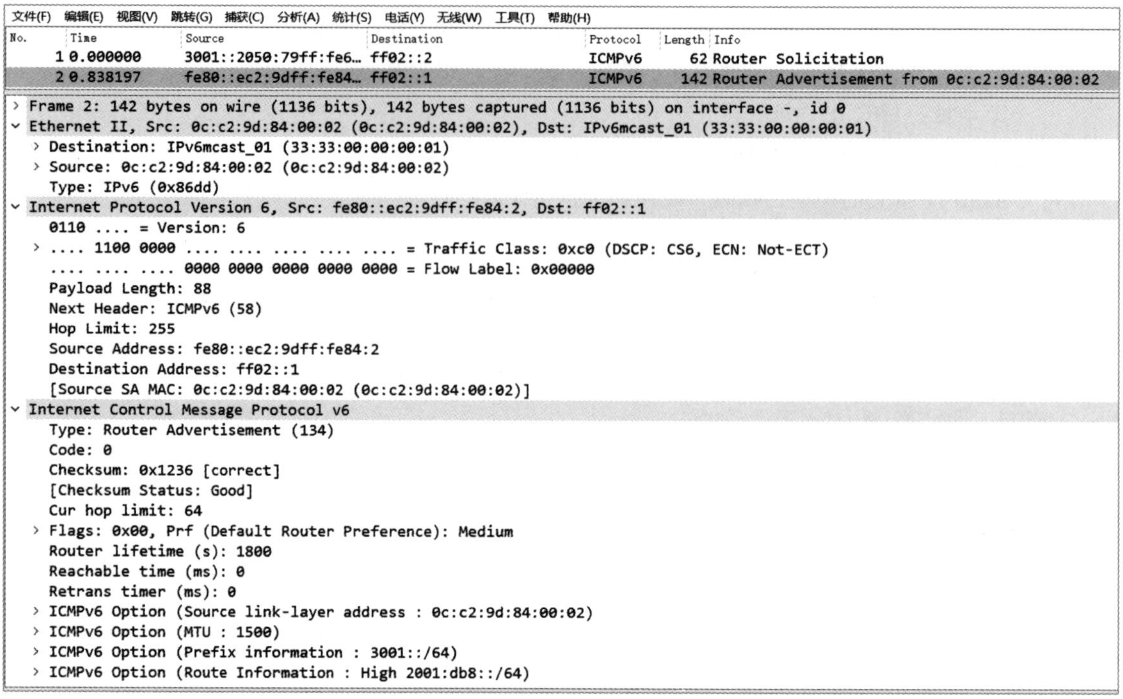

```
文件(F) 编辑(E) 视图(V) 跳转(G) 捕获(C) 分析(A) 统计(S) 电话(Y) 无线(W) 工具(T) 帮助(H)
No.     Time              Source                       Destination                   Protocol    Length  Info
   1    0.000000          3001::2050:79ff:fe6…         ff02::2                       ICMPv6          62  Router Solicitation
   2    0.838197          fe80::ec2:9dff:fe84…         ff02::1                       ICMPv6         142  Router Advertisement from 0c:c2:9d:84:00:02
> Frame 2: 142 bytes on wire (1136 bits), 142 bytes captured (1136 bits) on interface -, id 0
∨ Ethernet II, Src: 0c:c2:9d:84:00:02 (0c:c2:9d:84:00:02), Dst: IPv6mcast_01 (33:33:00:00:00:01)
   > Destination: IPv6mcast_01 (33:33:00:00:00:01)
   > Source: 0c:c2:9d:84:00:02 (0c:c2:9d:84:00:02)
     Type: IPv6 (0x86dd)
∨ Internet Protocol Version 6, Src: fe80::ec2:9dff:fe84:2, Dst: ff02::1
     0110 .... = Version: 6
   > .... 1100 0000 .... .... .... .... = Traffic Class: 0xc0 (DSCP: CS6, ECN: Not-ECT)
     .... .... .... 0000 0000 0000 0000 0000 = Flow Label: 0x00000
     Payload Length: 88
     Next Header: ICMPv6 (58)
     Hop Limit: 255
     Source Address: fe80::ec2:9dff:fe84:2
     Destination Address: ff02::1
     [Source SA MAC: 0c:c2:9d:84:00:02 (0c:c2:9d:84:00:02)]
∨ Internet Control Message Protocol v6
     Type: Router Advertisement (134)
     Code: 0
     Checksum: 0x1236 [correct]
     [Checksum Status: Good]
     Cur hop limit: 64
   > Flags: 0x00, Prf (Default Router Preference): Medium
     Router lifetime (s): 1800
     Reachable time (ms): 0
     Retrans timer (ms): 0
   > ICMPv6 Option (Source link-layer address : 0c:c2:9d:84:00:02)
   > ICMPv6 Option (MTU : 1500)
   > ICMPv6 Option (Prefix information : 3001::/64)
   > ICMPv6 Option (Route Information : High 2001:db8::/64)
```

图 8-3　RA 报文的结构

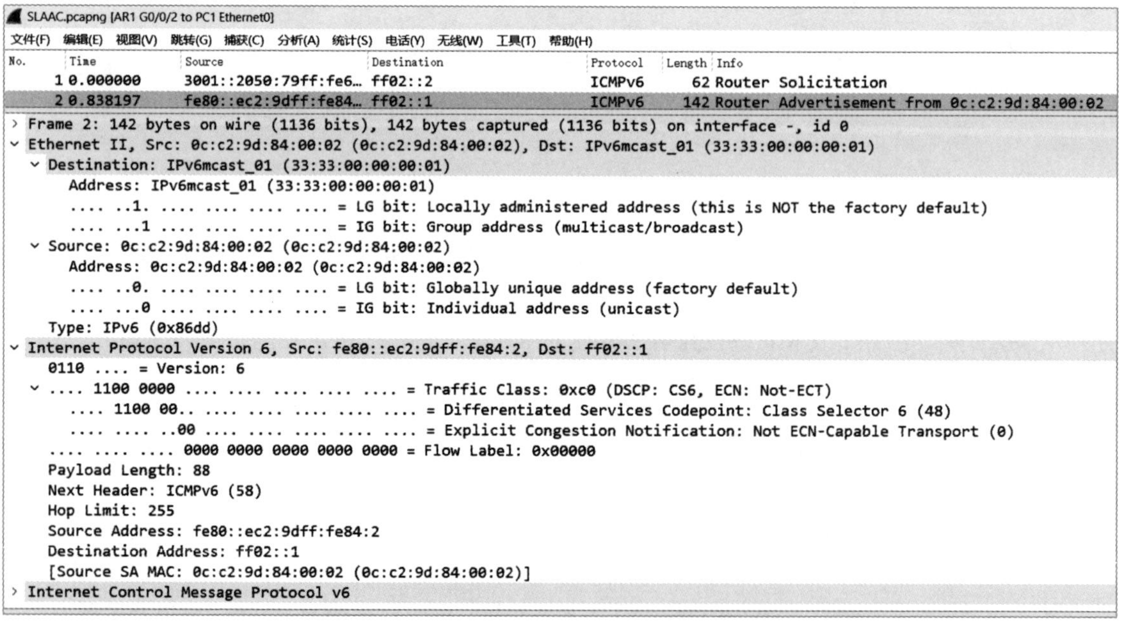

```
SLAAC.pcapng [AR1 G0/0/2 to PC1 Ethernet0]
文件(F) 编辑(E) 视图(V) 跳转(G) 捕获(C) 分析(A) 统计(S) 电话(Y) 无线(W) 工具(T) 帮助(H)
No.     Time              Source                       Destination                   Protocol    Length  Info
   1    0.000000          3001::2050:79ff:fe6…         ff02::2                       ICMPv6          62  Router Solicitation
   2    0.838197          fe80::ec2:9dff:fe84…         ff02::1                       ICMPv6         142  Router Advertisement from 0c:c2:9d:84:00:02
> Frame 2: 142 bytes on wire (1136 bits), 142 bytes captured (1136 bits) on interface -, id 0
∨ Ethernet II, Src: 0c:c2:9d:84:00:02 (0c:c2:9d:84:00:02), Dst: IPv6mcast_01 (33:33:00:00:00:01)
   ∨ Destination: IPv6mcast_01 (33:33:00:00:00:01)
       Address: IPv6mcast_01 (33:33:00:00:00:01)
       .... ..1. .... .... .... .... = LG bit: Locally administered address (this is NOT the factory default)
       .... ...1 .... .... .... .... = IG bit: Group address (multicast/broadcast)
   ∨ Source: 0c:c2:9d:84:00:02 (0c:c2:9d:84:00:02)
       Address: 0c:c2:9d:84:00:02 (0c:c2:9d:84:00:02)
       .... ..0. .... .... .... .... = LG bit: Globally unique address (factory default)
       .... ...0 .... .... .... .... = IG bit: Individual address (unicast)
     Type: IPv6 (0x86dd)
∨ Internet Protocol Version 6, Src: fe80::ec2:9dff:fe84:2, Dst: ff02::1
     0110 .... = Version: 6
   ∨ .... 1100 0000 .... .... .... .... = Traffic Class: 0xc0 (DSCP: CS6, ECN: Not-ECT)
     .... 1100 00.. .... .... .... .... = Differentiated Services Codepoint: Class Selector 6 (48)
     .... .... ..00 .... .... .... .... = Explicit Congestion Notification: Not ECN-Capable Transport (0)
     .... .... .... 0000 0000 0000 0000 0000 = Flow Label: 0x00000
     Payload Length: 88
     Next Header: ICMPv6 (58)
     Hop Limit: 255
     Source Address: fe80::ec2:9dff:fe84:2
     Destination Address: ff02::1
     [Source SA MAC: 0c:c2:9d:84:00:02 (0c:c2:9d:84:00:02)]
> Internet Control Message Protocol v6
```

图 8-4　RA 报文的 MAC 层和 IP 层封装

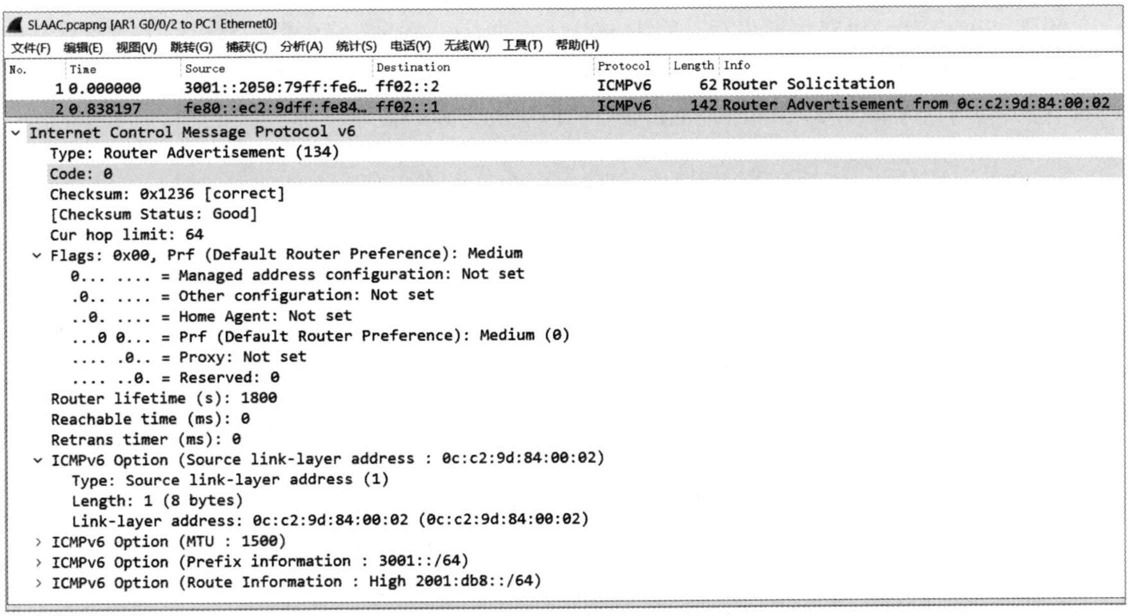

第 8 章 IPv6 无状态地址自动配置

```
SLAAC.pcapng [AR1 G0/0/2 to PC1 Ethernet0]
文件(F) 编辑(E) 视图(V) 跳转(G) 捕获(C) 分析(A) 统计(S) 电话(Y) 无线(W) 工具(T) 帮助(H)
No.    Time         Source                  Destination        Protocol   Length Info
  1    0.000000     3001::2050:79ff:fe6...  ff02::2            ICMPv6      62   Router Solicitation
  2    0.838197     fe80::ec2:9dff:fe84...  ff02::1            ICMPv6     142   Router Advertisement from 0c:c2:9d:84:00:02

Internet Control Message Protocol v6
  Type: Router Advertisement (134)
  Code: 0
  Checksum: 0x1236 [correct]
  [Checksum Status: Good]
  Cur hop limit: 64
  Flags: 0x00, Prf (Default Router Preference): Medium
    0... .... = Managed address configuration: Not set
    .0.. .... = Other configuration: Not set
    ..0. .... = Home Agent: Not set
    ...0 0... = Prf (Default Router Preference): Medium (0)
    .... .0.. = Proxy: Not set
    .... ..00 = Reserved: 0
  Router lifetime (s): 1800
  Reachable time (ms): 0
  Retrans timer (ms): 0
  ICMPv6 Option (Source link-layer address : 0c:c2:9d:84:00:02)
    Type: Source link-layer address (1)
    Length: 1 (8 bytes)
    Link-layer address: 0c:c2:9d:84:00:02 (0c:c2:9d:84:00:02)
  > ICMPv6 Option (MTU : 1500)
  > ICMPv6 Option (Prefix information : 3001::/64)
  > ICMPv6 Option (Route Information : High 2001:db8::/64)
```

图 8-5 RA 报文的 ICMPv6 封装（1）

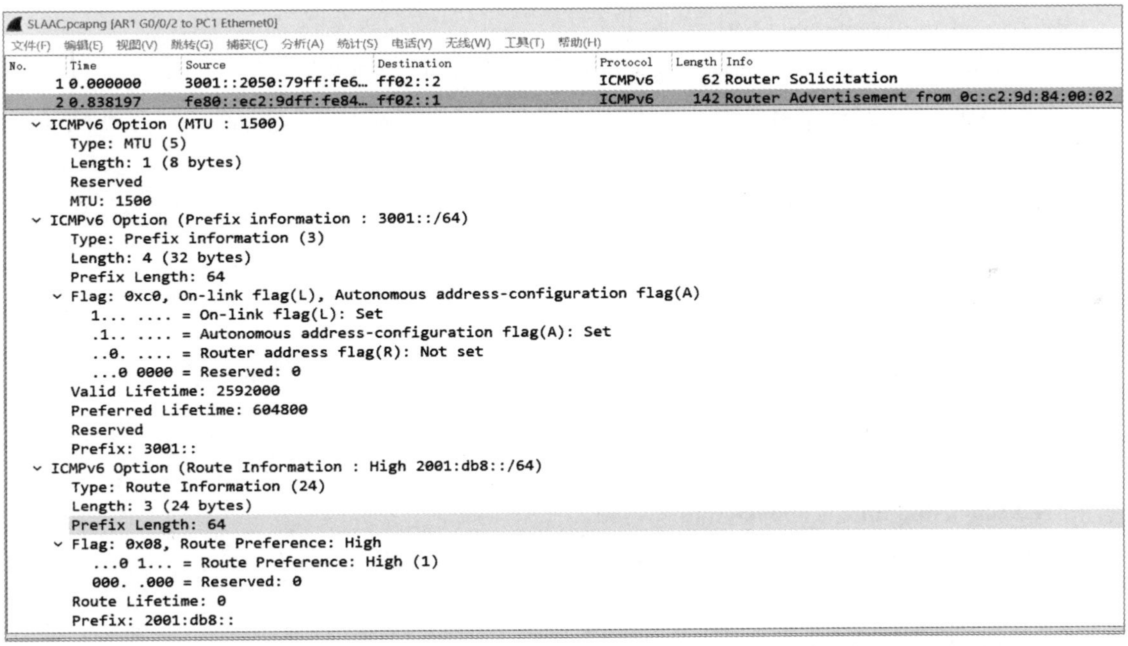

```
SLAAC.pcapng [AR1 G0/0/2 to PC1 Ethernet0]
文件(F) 编辑(E) 视图(V) 跳转(G) 捕获(C) 分析(A) 统计(S) 电话(Y) 无线(W) 工具(T) 帮助(H)
No.    Time         Source                  Destination        Protocol   Length Info
  1    0.000000     3001::2050:79ff:fe6...  ff02::2            ICMPv6      62   Router Solicitation
  2    0.838197     fe80::ec2:9dff:fe84...  ff02::1            ICMPv6     142   Router Advertisement from 0c:c2:9d:84:00:02

ICMPv6 Option (MTU : 1500)
  Type: MTU (5)
  Length: 1 (8 bytes)
  Reserved
  MTU: 1500
ICMPv6 Option (Prefix information : 3001::/64)
  Type: Prefix information (3)
  Length: 4 (32 bytes)
  Prefix Length: 64
  Flag: 0xc0, On-link flag(L), Autonomous address-configuration flag(A)
    1... .... = On-link flag(L): Set
    .1.. .... = Autonomous address-configuration flag(A): Set
    ..0. .... = Router address flag(R): Not set
    ...0 0000 = Reserved: 0
  Valid Lifetime: 2592000
  Preferred Lifetime: 604800
  Reserved
  Prefix: 3001::
ICMPv6 Option (Route Information : High 2001:db8::/64)
  Type: Route Information (24)
  Length: 3 (24 bytes)
  Prefix Length: 64
  Flag: 0x08, Route Preference: High
    ...0 1... = Route Preference: High (1)
    000. .000 = Reserved: 0
  Route Lifetime: 0
  Prefix: 2001:db8::
```

图 8-6 RA 报文的 ICMPv6 封装（2）

8）改变 RA 报文中相关字段的值

（1）将当前跳数限制字段"Cur hop limit"的值由默认值 64 更改为 32。

[cxyR1-GigabitEthernet0/0/2] ipv6 nd ra hop-limit 32

（2）将路由器生存时间字段"Router lifetime"的值更改为 900s。

[cxyR1-GigabitEthernet0/0/2] ipv6 nd ra router-lifetime 900

设置 G0/0/2 接口上路由器生存时间为 900s。在默认情况下，路由器生存时间为 RA 报文的最大发布间隔时间的 3 倍，即 600×3=1800（s）。

（3）将可达时间字段"Reachable time"的值更改为 600s。

[cxyR1-GigabitEthernet0/0/2] ipv6 nd nud reachable-time 600000   // 此命令中时间单位为 ms

设置 G0/0/2 接口上 RA 报文的可达时间（Reachable time）为 600s。

在默认情况下，RA 报文的可达时间未设置，该字段的值为 0，表示主机收到此 RA 报文后，将自身的可达时间设置为默认值 1200000ms。

（4）将重传定时器字段"Retrans timer"的值更改为 2s。

[cxyR1-GigabitEthernet0/0/2] ipv6 nd ns retrans-timer 2000   // 此命令中时间单位为 ms

设置 G0/0/2 接口上 RA 报文的"Retrans timer"的值为 2000ms。默认值为 1000ms。

（5）设置完成后，通过命令"display ipv6 interface g0/0/2"查看结果。

[cxyR1-GigabitEthernet0/0/2]display ipv6 interface g0/0/2

GigabitEthernet0/0/2 current state : UP

IPv6 protocol current state : UP

IPv6 is enabled, link-local address is FE80::EC2:9DFF:FE84:2

  Global unicast address(es):

    3001::1, subnet is 3001::/64

  Joined group address(es):

    FF02::1:FF00:1

    FF02::2

    FF02::1

    FF02::1:FF84:2

  MTU is 1500 bytes

  ND DAD is enabled, number of DAD attempts: 1

  ND **reachable time** is **600000** milliseconds

  ND **retransmit interval** is **2000** milliseconds

  ND stale time is 1200 seconds

  ND **advertised reachable time** is **600000** milliseconds

  ND **advertised retransmit interval** is **2000** milliseconds

  ND router advertisement max interval 600 seconds, min interval 200 seconds

  ND **router advertisements live** for **900** seconds

  ND **router advertisements hop-limit 32**

  ND default router preference medium

  Hosts use stateless autoconfig for addresses

4. 链路本地范围内有两台路由器启用发送 RA 报文功能，物理机的 VMnet8 接口和 R3 路由器的接口通过 SLAAC 方式获取前缀信息，生成全局 IPv6 地址

1）配置路由器 R1 接口 G0/0/1 启用发送 RA 报文功能

```
[cxyR1] interface g0/0/1
[cxyR1-GigabitEthernet0/0/1] ipv6 enable
// 手动配置 G0/0/1 接口的链路本地和全球单播 IP 地址
[cxyR1-GigabitEthernet0/0/1] ipv6 address fe80::1 link-local
[cxyR1-GigabitEthernet0/0/1] ipv6 address 2000::1/64
// 启用发送 RA 报文功能
[cxyR1-GigabitEthernet0/0/1] undo ipv6 nd ra halt
```

2）路由器 R3 的 G0/0/1 接口通过 SLAAC 方式获得全局 IPv6 地址

（1）从路由器 R1 处获得前缀地址并生成全球 IPv6 地址。

```
<Huawei>sys
Enter system view, return user view with Ctrl+Z
[Huawei]sys
[Huawei]sysname cxyR3    // 可将 cxy 改为你的名字拼音首字母
[cxyR3] ipv6
[cxyR3] interface g0/0/1
[cxyR3-GigabitEthernet0/0/1] ipv6 enable
// 接口通过 SLAAC 方式获取全球单播地址，并学习默认路由
[cxyR3-GigabitEthernet0/0/1] ipv6 address auto global default
```

（2）查看路由器 R3 的 G0/0/1 接口信息。

```
<cxyR3>display ipv6 interface g0/0/1
GigabitEthernet0/0/1 current state : UP
IPv6 protocol current state : UP
IPv6 is enabled, link-local address is FE80::E9A:7DFF:FE7B:1
  Global unicast address(es):
    2000::E9A:7DFF:FE7B:1,
    subnet is 2000::/64 [SLAAC 2024-04-06 07:19:37 2592000S]
  Joined group address(es):
    FF02::1:FF7B:1
    FF02::2
    FF02::1
  MTU is 1500 bytes
  ND DAD is enabled, number of DAD attempts: 1
  ND reachable time is 30000 milliseconds
```

ND retransmit interval is 1000 milliseconds
ND stale time is 1200 seconds

可见，路由器 R3 的 G0/0/1 接口已通过 SLAAC 方式成功获得了一个 IPv6 地址前缀 2000::/64，并据此生成了一个全局 IPv6 地址 2000::E9A:7DFF:FE7B:1。请解释说明该全局 IPv6 地址是如何生成的。

（3）查看路由器 R3 的路由表。

[cxyR3]display ipv6 routing-table
Routing Table : Public
    Destinations : 5   Routes : 5

Destination	: ::	PrefixLength	: 0
NextHop	: FE80::1	Preference	: 63
Cost	: 0	Protocol	: Unr
RelayNextHop	: ::	TunnelID	: 0x0
Interface	: GigabitEthernet0/0/1	Flags	: D
Destination	: ::1	PrefixLength	: 128
NextHop	: ::1	Preference	: 0
Cost	: 0	Protocol	: Direct
RelayNextHop	: ::	TunnelID	: 0x0
Interface	: InLoopBack0	Flags	: D
Destination	: 2000::	PrefixLength	: 64
NextHop	: 2000::E9A:7DFF:FE7B:1	Preference	: 0
Cost	: 0	Protocol	: Direct
RelayNextHop	: ::	TunnelID	: 0x0
Interface	: GigabitEthernet0/0/1	Flags	: D
Destination	: 2000::E9A:7DFF:FE7B:1	PrefixLength	: 128
NextHop	: ::1	Preference	: 0
Cost	: 0	Protocol	: Direct
RelayNextHop	: ::	TunnelID	: 0x0
Interface	: GigabitEthernet0/0/1	Flags	: D
Destination	: FE80::	PrefixLength	: 10
NextHop	: ::	Preference	: 0
Cost	: 0	Protocol	: Direct
RelayNextHop	: ::	TunnelID	: 0x0
Interface	: NULL0	Flags	: D

可见，有一条默认路由 ::/0，其下一跳 IPv6 地址是 FE80::1。

3）物理机的 VMnet8 接口通过 SLAAC 方式获得全局 IPv6 地址

（1）关闭 Windows 临时 IPv6 地址。

以管理员权限，在命令提示符下执行下列命令，然后重启物理机的 VMnet8 接口。

```
C:\Windows\system32> netsh interface ipv6 set privacy state=disable
```

（2）查看物理机的 VMnet8 接口的信息。

```
C:\Windows\system32>ipconfig /all
```
以太网适配器 VMware Network Adapter VMnet8_ 192.168.241.1：

连接特定的 DNS 后缀	:
描述	: VMware Virtual Ethernet Adapter for VMnet8
物理地址	: 00-50-56-C0-00-08
DHCP 已启用	: 是
自动配置已启用	: 是
IPv6 地址	**: 2000::2419:17de:7b0b:793(首选)**
本地链接 IPv6 地址	: fe80::9be9:338c:ba9e:2d75%19(首选)
IPv4 地址	: 192.168.241.1(首选)
子网掩码	: 255.255.255.0
获得租约的时间	: 2024 年 4 月 6 日 10:30:24
租约过期的时间	: 2024 年 4 月 6 日 16:16:42
默认网关	**: fe80::1%19**
DHCP 服务器	: 192.168.241.254
DHCPv6 IAID	: 402673750
DHCPv6 客户端 DUID	: 00-01-00-01-2A-E9-44-54-98-8F-E0-61-18-2B
主 WINS 服务器	: 192.168.241.2
TCPIP 上的 NetBIOS	: 已启用

请问物理机 VMnet8 接口的 IPv6 地址和默认网关分别是什么？并解释原因。

4）配置路由器 R2 启用发送 RA 报文功能

（1）在路由器 R2 接口 G0/0/1 上启用发送 RA 报文功能。

```
<Huawei>system-view
Enter system view, return user view with Ctrl+Z.
[Huawei] sysname cxyR2          // 可将 cxy 改为你的名字拼音首字母
[cxyR2]
[cxyR2]user-interface console 0
[cxyR2-ui-console0] idle-timeout 0     // 配置 console 口不超时
Info: The timeout period is 0, the terminal will remain in the login state, leading to security risks.
```

[cxyR2] ipv6
[cxyR2] interface g0/0/1
[cxyR2-GigabitEthernet0/0/1] ipv6 enable
// 手动配置 G0/0/1 接口的链路本地和全球单播 IP 地址
[cxyR2-GigabitEthernet0/0/1] ipv6 address fe80::2 link-local
[cxyR2-GigabitEthernet0/0/1] ipv6 address 2000::2/64
// 启用发送 RA 报文功能
[cxyR2-GigabitEthernet0/0/1] undo ipv6 nd ra halt

注意：此时，在同一链路范围内有两台路由器 R1 和 R2 均开启了发送 RA 报文功能，路由器 R1、R2 的默认优先级均为 medium。

（2）查看路由器 R3 的路由表。

<cxyPC>display ipv6 routing-table
[cxyR3]display ipv6 routing-table
Routing Table : Public
　　　Destinations : 5　　**Routes : 6**
……

Destination	: ::	PrefixLength	: 0
NextHop	: FE80::1	Preference	: 63
Cost	: 0	Protocol	: Unr
RelayNextHop	: ::	TunnelID	: 0x0
Interface	: GigabitEthernet0/0/1	Flags	: D

Destination	: ::	PrefixLength	: 0
NextHop	: FE80::2	Preference	: 63
Cost	: 0	Protocol	: Unr
RelayNextHop	: ::	TunnelID	: 0x0
Interface	: GigabitEthernet0/0/1	Flags	: D

……

思考：此时，路由器 R3 的路由表中有几条默认路由？为什么？其下一跳（网关）IP 地址分别是什么？注意：路由器 R1 和 R2 的默认优先级均为 medium。

（3）查看物理机的 VMnet8 接口的信息。

C:\Windows\system32>ipconfig /all
以太网适配器 VMware Network Adapter VMnet8_ 192.168.241.1:

　　连接特定的 DNS 后缀　　　　　　　　：
　　描述　　　　　　　　　　　　　　　 ：VMware Virtual Ethernet Adapter for VMnet8
　　物理地址　　　　　　　　　　　　　 ：00-50-56-C0-00-08
　　DHCP 已启用　　　　　　　　　　　　：是

自动配置已启用	: 是
IPv6 地址	**: 2000::2419:17de:7b0b:793(首选)**
本地链接 IPv6 地址	: fe80::9be9:338c:ba9e:2d75%19(首选)
IPv4 地址	: 192.168.241.1(首选)
子网掩码	: 255.255.255.0
获得租约的时间	: 2024 年 4 月 6 日 10:35:24
租约过期的时间	: 2024 年 4 月 6 日 16:21:42
默认网关	**: fe80::1%19**
	: fe80::2%19
DHCP 服务器	: 192.168.241.254
DHCPv6 IAID	: 402673750
DHCPv6 客户端 DUID	: 00-01-00-01-2A-E9-44-54-98-8F-E0-61-18-2B
主 WINS 服务器	: 192.168.241.2
TCPIP 上的 NetBIOS	: 已启用

5）将路由器 R2 的默认优先级由 medium（默认值）更改为 high

[cxyR2-GigabitEthernet0/0/1] ipv6 nd ra preference high

注意：此时，在同一链路范围内路由器 R1 和 R2 均启用了发送 RA 报文功能，但 R2 的默认优先级为 high，而 R1 的默认优先级为 medium。可以通过抓包分析，验证 R2 的默认优先级由 medium 更改为 high。

（1）查看路由器 R3 的路由表。

```
<cxyPC>display ipv6 routing-table
[cxyR3]display ipv6 routing-table
Routing Table : Public
    Destinations : 5      Routes : 5

Destination        : ::                    PrefixLength    : 0
NextHop            : FE80::2               Preference      : 63
Cost               : 0                     Protocol        : Unr
RelayNextHop       : ::                    TunnelID        : 0x0
Interface          : GigabitEthernet0/0/1  Flags           : D
……
```

此时，只有一条默认路由。

（2）查看物理机的 VMnet8 接口的信息。

C:\Windows\system32>ipconfig /all
以太网适配器 VMware Network Adapter VMnet8_ 192.168.241.1:

　　连接特定的 DNS 后缀　　　　　　　　　　　　:

描述	: VMware Virtual Ethernet Adapter for VMnet8
物理地址	: 00-50-56-C0-00-08
DHCP 已启用	: 是
自动配置已启用	: 是
IPv6 地址	**: 2000::2419:17de:7b0b:793(首选)**
本地链接 IPv6 地址	: fe80::9be9:338c:ba9e:2d75%19(首选)
IPv4 地址	: 192.168.241.1(首选)
子网掩码	: 255.255.255.0
获得租约的时间	: 2024 年 4 月 6 日 10:35:24
租约过期的时间	: 2024 年 4 月 6 日 16:21:42
默认网关	**: fe80::1%19**
	: fe80::2%19
DHCP 服务器	: 192.168.241.254
DHCPv6 IAID	: 402673750
DHCPv6 客户端 DUID	: 00-01-00-01-2A-E9-44-54-98-8F-E0-61-18-2B
主 WINS 服务器	: 192.168.241.2
TCPIP 上的 NetBIOS	: 已启用

第 9 章
DHCPv6——地址自动分配

9.1 组网需求

如图 9-1 所示，路由器 R1、路由器 R2 和 Cloud1（物理机）在同一条链路上。通过部署 R2 作为 DHCPv6 服务器、R1 和 Cloud1 作为 DHCPv6 客户端，实现 R1 和 Cloud1 通过 DHCPv6 服务器获取 IPv6 地址和 DNS 服务器地址的需求，具体如下。

（1）在 R2 上，配置 DHCPv6 服务器，实现为 DHCPv6 客户端动态分配 IPv6 地址和 DNS 服务器地址，具体包括配置接口的 IPv6 地址、配置 IPv6 地址池、启用接口的 DHCPv6 服务器功能。

（2）在 R1 上，配置 DHCPv6 客户端功能，实现通过 DHCPv6 服务器动态获取 IPv6 地址和 DNS 服务器地址，具体包括配置接口的 IPv6 功能、启用接口的 DHCPv6 客户端功能。

（3）在 Cloud1（物理机）的 VMnet8 接口上进行配置，实现通过 DHCPv6 动态获取 IPv6 地址和 DNS 服务器地址。

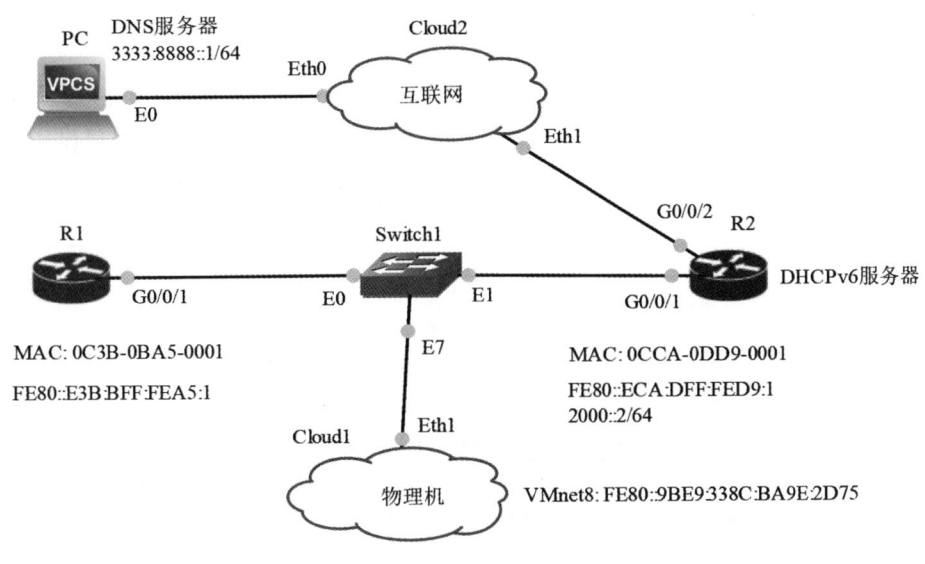

图 9-1 网络拓扑

当物理机的接口首次获得地址时，DHCPv6 客户端与服务器需要经过 4 次交互过程，这 4

次交互对应的报文分别是 Solicit 报文、Advertise 报文、Request 报文和 Reply 报文。

当接口进行有状态 IP 地址更新时，DHCPv6 客户端与服务器需要经历两次交互过程，这两次交互对应的报文分别是 Renew 报文和 Reply 报文。

当接口释放 IP 地址时，DHCPv6 客户端与服务器需要经历两次交互过程，这两次交互对应的报文分别是 Release 报文和 Reply 报文。

链路范围内的所有 DHCPv6 服务器的组播地址是 FF02::1:2，用于客户端和服务器之间的通信，所有 DHCPv6 服务器和中继代理都是该组播组的成员。DHCPv6 报文封装在 UDPv6 上，客户端侦听的 UDP 端口号是 546，服务器、中继代理侦听的 UDP 端口号是 547。

9.2 仿真实验

9.2.1 仿真环境

仿真环境所需要的硬件、软件建议至少满足以下要求。

- 物理机（PC）的处理器主频为 1.6GHz；
- 内存 RAM 大小为 32.0GB；
- 物理机的操作系统为 Windows 10；
- 虚拟机为 VMware Workstation 16 Pro；
- 网络模拟器选用 GNS3 2.2.31；
- 交换机为 Ethernet Switch；
- 路由器为 Huawei AR1kV；
- PC 为 Virtual PC Simulator, Version 0.8.2。

9.2.2 实验设计

根据网络拓扑（见图 9-1）完成网络设备的选型：路由器 R1 和 R2 均为华为 AR 系列路由器，PC 通过 Cloud2 的 Eth0 接口接入 GNS3 网络，而物理机通过 Cloud1 的 Eth1 接口接入 GNS3 网络。根据前面的组网需求，本实验设计如下。

1. VLAN 设计

交换机无须创建新的 VLAN，所有接口均属于 VLAN1。

2. IP 地址规划

各设备接口的 IPv6 地址/前缀如表 9-1 所示。

表 9-1 各设备接口的 IPv6 地址/前缀

设备	接口	全局 IP 地址/前缀	链路本地 IP 地址/前缀
R1	G0/0/1	通过 DHCPv6 获得 IP 地址/前缀	auto
R2	G0/0/1	2000::2/64	auto
物理机	VMnet8	通过 DHCPv6 获得 IP 地址/前缀	auto
PC	E0	3333:8888::1/64	auto

3. RA 报文参数设置

设置 RA 报文中不包含"前缀信息"（Prefix information）选项。默认情况下，RA 报文中携带"前缀信息"选项，当接口收到 RA 报文后，将提取该选项的值作为前缀，然后自动地为本接口生成一个 IPv6 全局地址。

设置 RA 报文的 M 标志位和 O 标志位，使得 DHCPv6 客户端通过 DHCPv6 方式获取 IPv6 地址和其他网络参数。

4. R2、R1 和物理机配置

在 R2 上，配置 DHCPv6 服务器，主要包括 IPv6 地址池、IPv6 地址与客户端 DUID 静态绑定、续租时间、重绑定时间和冲突地址老化时间。在 R1 和物理机上，配置 DHCPv6 客户端，主要包括 IPv6 链路本地地址、通过 DHCP 方式自动获得全局地址、通过 RA 报文获得下一跳为网关的默认路由。

9.2.3 操作步骤

下面给出本次实验的具体操作。

1. 按照图 9-1 所示的网络拓扑，在 GNS3 中新建一个空白的工程

两台路由器选用华为的 AR 系列路由器，具体型号为"Huawei AR1kV V300R021C00SPC100T"。

模拟 DNS 服务器的 PC 通过 Cloud2 的 Eth0 接口（对应虚拟网卡的 VMware Network Adapter VMnet1 网络，IP 地址为 192.168.148.1/24）接入 GNS3 网络中。

物理机通过 Cloud1 的 Eth1 接口（对应虚拟网卡的 VMware Network Adapter VMnet8 网络，IP 地址为 192.168.241.1/24）接入 GNS3 网络中。

2. 在 R2 上，配置 DHCPv6 服务器

1）配置接口的 IPv6 地址

```
<HUAWEI> system-view
[HUAWEI] sysname cxyR2            // 可将其中的 cxy 改为你的名字拼音首字母
[cxyR2] ipv6
[cxyR2] interface g0/0/1
[cxyR2-GigabitEthernet0/0/1] ipv6 enable
[cxyR2-GigabitEthernet0/0/1] ipv6 address 2000::2/64
[SwitchA-Vlanif100] quit
```

2）配置 IPv6 地址池

（1）创建名为"cxypool"的 IPv6 地址池。

```
[cxyR2] dhcpv6 pool cxypool        // 可将其中的 cxy 改为你的名字拼音首字母
```

（2）配置网络前缀和生命周期。

```
[cxyR2-dhcpv6-pool-cxypool] address prefix 2000::/64 life-time 3600 1800
```

（3）设置不参与自动分配的 IPv6 地址列表。

```
[cxyR2-dhcpv6-pool-cxypool] excluded-address 2000::1FF to 2000::2FF
```

通常将已分配给路由器接口的 IP 地址、已分配或未来可能要分配给服务器的 IP 地址加入不参与自动分配的 IPv6 地址列表中。

（4）设置 DNS 服务器地址和 DNS 域名后缀。

可以分别用命令"dns-server"和"dns-domain-name"为名为"cxypool"的地址池配置网络参数信息：DNS 服务器地址和 DNS 域名后缀。在默认情况下，IPv6 地址池未配置 DNS 服务器地址和 DNS 域名后缀。

配置 DNS 服务器地址为 3333:8888::1、DNS 域名后缀为 cxy.com：

```
[cxyR2-dhcpv6-pool-cxypool] dns-server 3333:8888::1
[cxyR2-dhcpv6-pool-cxypool] dns-domain-name cxy.com
```

（5）分配固定的 IPv6 地址给特定的客户端。

将 IPv6 地址与客户端 DUID（DHCP Unique Identifier，DHCP 唯一 ID）进行静态绑定，就可为指定的客户端分配固定的 IPv6 地址。所谓 DUID，是唯一标识一台 DHCPv6 设备的标识符，该设备可能是 DHCP 客户端、DHCP 中继或者 DHCP 服务器。

现要求给物理机 VMnet8 接口分配固定的 IPv6 地址 2000::1FF，则需要事先知道物理机的 DUID。对于 Windows 主机，可用"ipconfig /all"命令获得设备的 DUID。例如：

```
C:\Users\lll>ipconfig /all
以太网适配器 VMware Network Adapter VMnet8:

   连接特定的 DNS 后缀       . . . . . . . :
   描述.  . . . . . . . . . . . . . . : VMware Virtual Ethernet Adapter for VMnet8
   物理地址. . . . . . . . . . . . . : 00-50-56-C0-00-08
   DHCP 已启用 . . . . . . . . . . . : 是
   自动配置已启用. . . . . . . . . . : 是
   本地链接 IPv6 地址. . . . . . . . : fe80::9be9:338c:ba9e:2d75%19(首选)
   IPv4 地址 . . . . . . . . . . . . : 192.168.241.1(首选)
   子网掩码  . . . . . . . . . . . . : 255.255.255.0
   获得租约的时间  . . . . . . . . . : 2024 年 4 月 6 日 10:30:24
   租约过期的时间  . . . . . . . . . : 2024 年 4 月 7 日 15:21:23
   默认网关. . . . . . . . . . . . . : fe80::eca:dff:fed9:1%19
   DHCP 服务器 . . . . . . . . . . . : 192.168.241.254
   DHCPv6 IAID . . . . . . . . . . . : 402673750
   DHCPv6 客户端 DUID. . . . . . . . : 00-01-00-01-2A-E9-44-54-98-8F-E0-61-18-2B
   主 WINS 服务器  . . . . . . . . . : 192.168.241.2
   TCPIP 上的 NetBIOS  . . . . . . . : 已启用
```

可见，本设备的 DHCPv6 客户端 DUID 值为 000100012AE94454988FE061182B。值得一提的是：此时物理机的 VMnet8 接口虽然没有全局 IPv6 地址，但已经有链路本地 IPv6 地址和默认网关了。获得 DUID 后，可以用下面的命令将 IPv6 地址 2000::AAA 与 DUID 为 000100012AE94454988FE061182B 的客户端进行静态绑定：

[cxyR2-dhcpv6-pool-cxypool] static-bind address 2000:: AAA duid 000100012AE94454988FE061182B

（6）配置续租时间和重绑定时间。

在默认情况下，DHCPv6 的续租时间是首选生命周期（Preferred lifetime）的 50%，重绑定时间是首选生命周期的 80%。可以用下面的命令将续租时间和重绑定时间分别设置为首选生命周期的 60%和 90%：

[cxyR2-dhcpv6-pool-cxypool] renew-time-percent 60 rebind-time-percent 90

（7）配置无状态方式下地址池的网络参数信息刷新时间。

在默认情况下，无状态 DHCPv6 的信息刷新时间为 1 天，即 86400s。可以用下面的命令将无状态 DHCPv6 的网络参数信息刷新时间设置为 2 天：

[cxyR2-dhcpv6-pool-cxypool] information-refresh 172800

（8）配置地址池中冲突地址的老化时间。

在默认情况下，地址池中冲突地址的老化时间是 2 天，即 172800s。用下面的命令将冲突地址的老化时间配置为 86400s。

[cxyR2-dhcpv6-pool-cxypool] conflict-address expire-time 86400

（9）查看 cxypool 地址池信息。

```
<cxyR2>display dhcpv6 pool cxypool
DHCPv6 pool: cxypool
 Address prefix: 2000::/64
   Lifetime valid 3600 seconds, preferred 1800 seconds
   2 in use, 0 conflicts
 static-bind address 2000::AAA duid 000100012AE94454988FE061182B
   Lifetime valid 172800 seconds, preferred 86400 seconds
 1 static addresses
 excluded-address 2000::1FF to 2000::2FF
 257 excluded addresses
 Information refresh time: 172800
 DNS server address: 3333:8888::1
 Domain name: cxy.com
 conflict-address expire-time: 86400
 renew-time-percent : 60
 rebind-time-percent : 90
 Active normal clients: 2
 Logging : Disable
```

3）在接口上启用 DHCPv6 服务器功能

[cxyR2] **dhcp enable**
[cxyR2] **interface g0/0/1**
[cxyR2-GigabitEthernet0/0/1] **dhcpv6 server** cxypool

4）设置 RA 报文中的相应字段值

（1）在接口上启用发送 RA 报文功能。

[cxyR2-GigabitEthernet0/0/1] **undo ipv6 nd ra halt**

（2）设置 RA 报文中不含"前缀信息"选项。

设置 RA 报文中不包含"前缀信息"（Prefix information）选项。默认情况下，RA 报文中携带了"前缀信息"选项，当接口收到 RA 报文后，将提取该选项的值作为前缀，然后自动地为本接口生成一个 IPv6 全局地址。

[cxyR2-GigabitEthernet0/0/1] ipv6 nd ra prefix default no-advertise

下面以物理机获得的 IPv6 地址来说明"前缀信息"选项的作用。当 RA 报文中携带"前缀信息"选项时，抓包得到的"前缀信息"选项的内容如下。

```
ICMPv6 Option (Prefix information: 2000::/64)
   Type: Prefix information (3)
   Length: 4 (32 bytes)
   Prefix Length: 64
   Flag: 0xc0, On-link flag(L), Autonomous address-configuration flag(A)
     1... .... = On-link flag(L): Set
     .1.. .... = Autonomous address-configuration flag(A): Set
     ..0. .... = Router address flag(R): Not set
     ...0 0000 = Reserved: 0
   Valid Lifetime: 2592000
   Preferred Lifetime: 604800
   Reserved
   Prefix: 2000::
```

注意：该选项中的前缀长度是 64，前缀值是 2000::。当物理机的 VMnet8 收到此 RA 报文后，将据此自动生成一个前缀为 2000:: 的 IPv6 地址。

```
C:\Users\lll>ipconfig /all
以太网适配器 VMware Network Adapter VMnet8:

   连接特定的 DNS 后缀  . . . . . . . : cxy.com
   描述. . . . . . . . . . . . . . . : VMware Virtual Ethernet Adapter for VMnet8
   物理地址. . . . . . . . . . . . . : 00-50-56-C0-00-08
```

DHCP 已启用	: 是
自动配置已启用	: 是
IPv6 地址	**: 2000::aaa(首选)**
获得租约的时间	: 2024 年 4 月 7 日 10:05:25
租约过期的时间	: 2024 年 4 月 9 日 10:05:25
IPv6 地址	**: 2000::2419:17de:7b0b:793(首选)**
本地链接 IPv6 地址	: fe80::9be9:338c:ba9e:2d75%19(首选)

可以看到，物理机的 VMnet8 接口现在有两个全局 IPv6 地址：2000::AAA 和 2000::2419:17DE:7B0B:793，前者从 DHCPv6 服务器获得；后者是从 RA 报文的"前缀信息"选项得到 64 位前缀，然后通过随机的方式生成后 64 位，最终得到的 128 位全局 IPv6 地址。

当 RA 报文不携带"前缀信息"选项时，物理机的 VMnet8 接口将只有一个从 DHCPv6 服务器获得的全局 IPv6 地址：2000::AAA。

（3）设置 RA 报文的 M 标志位和 O 标志位。

设置 RA 报文的 M 标志位和 O 标志位的值为 1，其作用是告诉 DHCPv6 客户端通过 DHCPv6 方式获取 IPv6 地址和其他网络参数。

[cxyR2-GigabitEthernet0/0/1] ipv6 nd autoconfig managed-address-flag
[cxyR2-GigabitEthernet0/0/1] ipv6 nd autoconfig other-flag

3. 在 R1 上，配置 DHCPv6 客户端

1）配置接口的 IPv6 链路本地地址

```
<HUAWEI> system-view
[HUAWEI] sysname cxyR1              // 可将其中的 cxy 改为你的名字拼音首字母
[cxyR1] ipv6
[cxyR1] interface g0/0/1
[cxyR1-GigabitEthernet0/0/1] ipv6 enable
[cxyR1-GigabitEthernet0/0/1] ipv6 address auto link-local
```

2）设置接口的全局地址为通过 DHCP 方式自动获得

[cxyR1-GigabitEthernet0/0/1] ipv6 address auto dhcp

启用接口的 DHCPv6 客户端功能，该接口将从 DHCPv6 服务器通过有状态方式自动获取 IPv6 全局地址，同时获得其他网络配置参数，如 DNS 服务器 IP 地址等信息。

3）设置接口通过 RA 报文获得网关的默认路由

[cxyR1-GigabitEthernet0/0/1] ipv6 address auto global default

设置 DHCPv6 客户端通过 RA 报文获得 IPv6 网关的默认路由，若不进行此设置，R1 的路由表中没有到达 R2 的默认路由。

4）查看 DHCPv6 通过有状态方式获取的 IPv6 全局地址

（1）查看 DHCPv6 客户端的配置信息。

```
[cxyR1]display dhcpv6 client
GigabitEthernet0/0/1 is in stateful DHCPv6 client mode
Stateful DHCPv6 client is in BOUND state
Preferred server DUID         : 000300010CCA0DD90000
  Reachable via address       : FE80::ECA:DFF:FED9:1
IA NA IA ID 0x00000041 T1 918 T2 1458
  Obtained         : 2024-04-07 20:32:56
  Renews           : 2024-04-07 20:48:14
  Rebinds          : 2024-04-07 20:57:14
  Address          : 2000::1
    Lifetime valid 3600 seconds, preferred 1800 seconds
    Expires at 2024-04-07 21:32:56 (2953 seconds left)
DNS server         : 3333:8888::1
```

（2）查看 DHCPv6 客户端上已经生成的到 IPv6 网关的默认路由。

```
<cxyR1>display ipv6 routing-table
Routing Table : Public
      Destinations : 4    Routes : 4

Destination    : ::                        PrefixLength  : 0
NextHop        : FE80::ECA:DFF:FED9:1      Preference    : 64
Cost           : 0                         Protocol      : Unr
RelayNextHop   : ::                        TunnelID      : 0x0
Interface      : GigabitEthernet0/0/1      Flags         : D

Destination    : ::1                       PrefixLength  : 128
NextHop        : ::1                       Preference    : 0
Cost           : 0                         Protocol      : Direct
RelayNextHop   : ::                        TunnelID      : 0x0
Interface      : InLoopBack0               Flags         : D

Destination    : 2000::1                   PrefixLength  : 128
NextHop        : ::1                       Preference    : 0
Cost           : 0                         Protocol      : Direct
RelayNextHop   : ::                        TunnelID      : 0x0
Interface      : GigabitEthernet0/0/1      Flags         : D

Destination    : FE80::                    PrefixLength  : 10
NextHop        : ::                        Preference    : 0
```

Cost	: 0	Protocol	: Direct
RelayNextHop	: ::	TunnelID	: 0x0
Interface	: NULL0	Flags	: D

4. 在物理机的 VMnet8 接口上，配置 DHCPv6 客户端

（1）在如图 9-2 的所示"VMware Network Adapter VMnet 8-192.168.241.1 属性"窗口中，选择"Internet 协议版本 6 (TCP/IPv6)"选项，然后单击"属性"按钮。

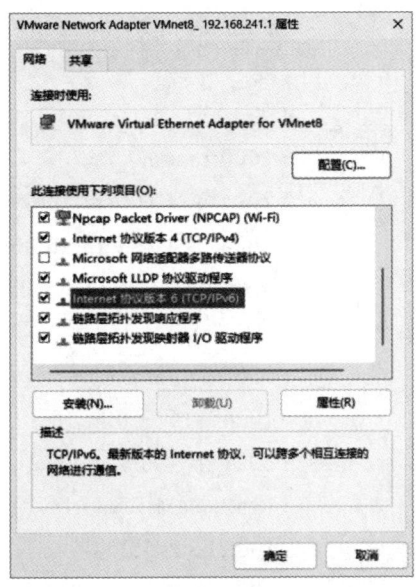

图 9-2　"VMware Network Adapter VMnet 8_192.168.241.1 属性"窗口

（2）在如图 9-3 所示的"Internet 协议版本 6(TCP/IPv6) 属性"窗口中，选中"自动获取 IPv6 地址"单选按钮和"自动获得 DNS 服务器地址"单选按钮，然后单击"确定"按钮。

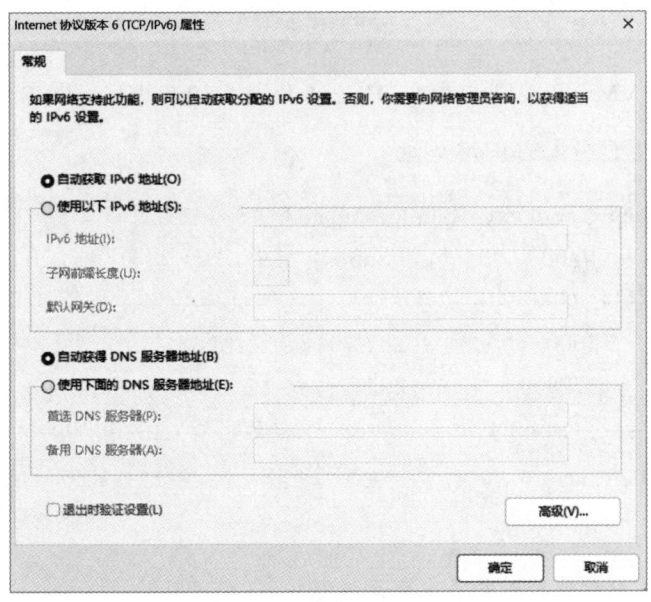

图 9-3　"Internet 协议版本 6(TCP/IPv6)属性"窗口

（3）查看物理机的 VMnet8 接口的信息。

```
C:\Users\lll>ipconfig /all
```
以太网适配器 VMware Network Adapter VMnet8:

连接特定的 DNS 后缀	: cxy.com
描述	: VMware Virtual Ethernet Adapter for VMnet8
物理地址	: 00-50-56-C0-00-08
DHCP 已启用	: 是
自动配置已启用	: 是
IPv6 地址	: 2000::aaa(首选)
获得租约的时间	: 2024 年 4 月 7 日 10:05:25
租约过期的时间	: 2024 年 4 月 9 日 10:05:25
本地链接 IPv6 地址	: fe80::9be9:338c:ba9e:2d75%19(首选)
IPv4 地址	: 192.168.241.1(首选)
子网掩码	: 255.255.255.0
获得租约的时间	: 2024 年 4 月 6 日 10:30:24
租约过期的时间	: 2024 年 4 月 7 日 15:21:23
默认网关	: fe80::eca:dff:fed9:1%19
DHCP 服务器	: 192.168.241.254
DHCPv6 IAID	: 402673750
DHCPv6 客户端 DUID	: 00-01-00-01-2A-E9-44-54-98-8F-E0-61-18-2B
DNS 服务器	: 3333:8888::1
主 WINS 服务器	: 192.168.241.2
TCPIP 上的 NetBIOS	: 已启用

可见，物理机的 VMnet8 接口已通过 DHCPv6 获得了 IPv6 地址、DNS 服务器地址等信息。

5. 在 R2 上，查看已分配的 IPv6 地址

```
[cxyR2]display dhcpv6 pool cxypool allocated address
Address          Valid       Expires                Left
-----------------------------------------------------------------
2000::1          3600        2024-04-07 22:18:55    3269
2000::AAA        172800      2024-04-09 17:33:00    158914
-----------------------------------------------------------------
Total : 2
```

6. 抓包分析

1）DHCPv6 首次分配地址时的 4 次交互过程

当接口首次执行命令"ipv6 address auto dhcp"进行有状态 IP 地址自动分配时，DHCPv6 客

户端与服务器需要经历 4 次交互过程，这 4 次交互对应的报文分别是 Solicit 报文（见图 9-4）、Advertise 报文（见图 9-5）、Request 报文（见图 9-6）和 Reply 报文（见图 9-7）。

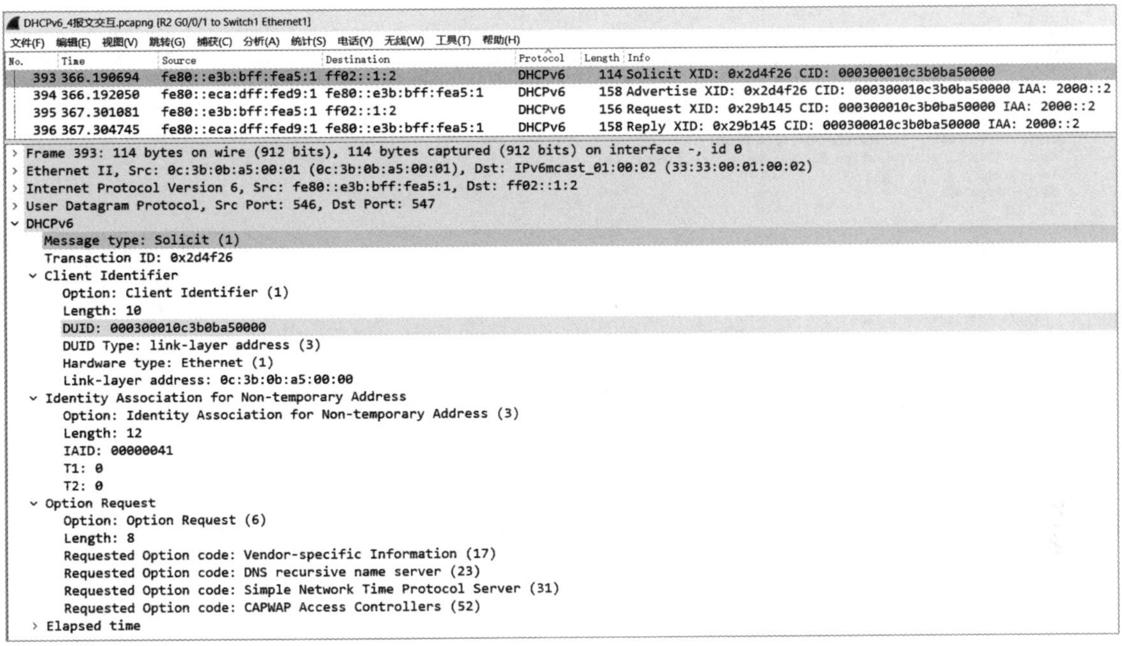

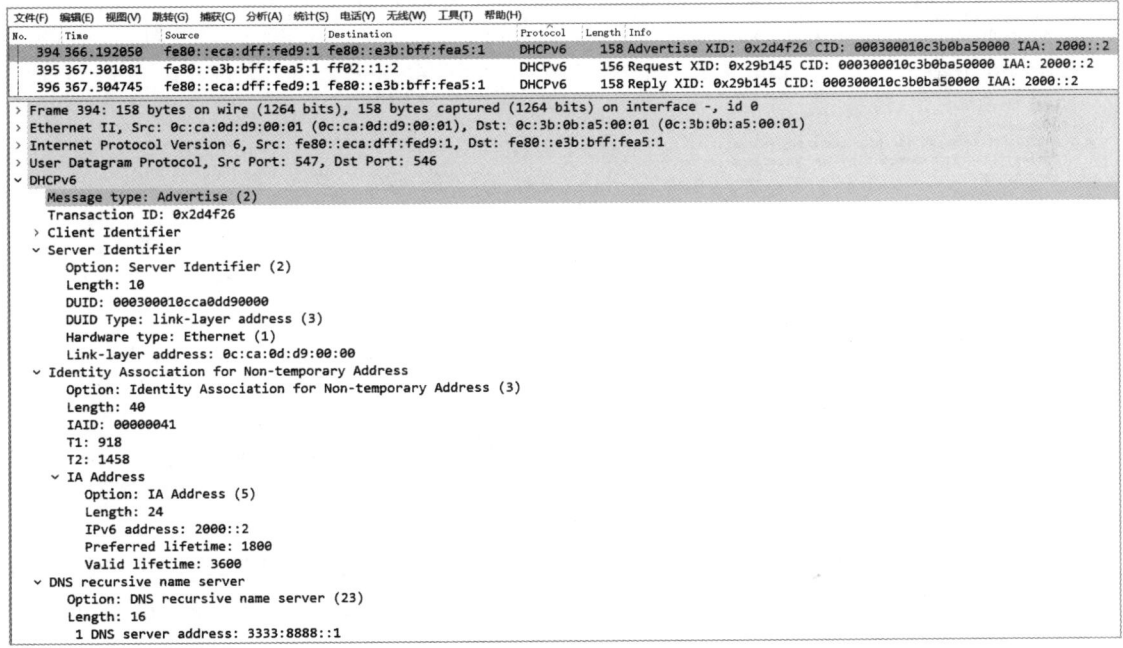

图 9-4　DHCPv6 首次分配地址时第 1 次交互对应的 Solicit 报文

图 9-5　DHCPv6 首次分配地址时第 2 次交互对应的 Advertise 报文

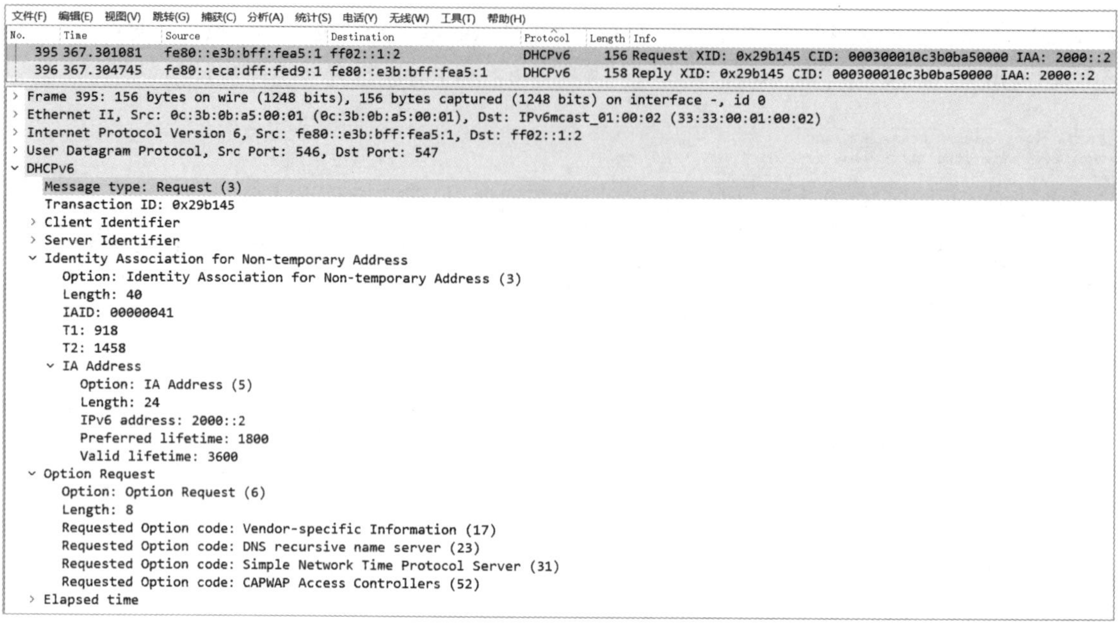

```
文件(F)  编辑(E)  视图(V)  跳转(G)  捕获(C)  分析(A)  统计(S)  电话(Y)  无线(W)  工具(T)  帮助(H)
No.    Time          Source                    Destination            Protocol  Length  Info
 395 367.301081   fe80::e3b:bff:fea5:1      ff02::1:2              DHCPv6    156 Request XID: 0x29b145 CID: 000300010c3b0ba50000 IAA: 2000::2
 396 367.304745   fe80::eca:dff:fed9:1      fe80::e3b:bff:fea5:1   DHCPv6    158 Reply XID: 0x29b145 CID: 000300010c3b0ba50000 IAA: 2000::2

> Frame 395: 156 bytes on wire (1248 bits), 156 bytes captured (1248 bits) on interface -, id 0
> Ethernet II, Src: 0c:3b:0b:a5:00:01 (0c:3b:0b:a5:00:01), Dst: IPv6mcast_01:00:02 (33:33:00:01:00:02)
> Internet Protocol Version 6, Src: fe80::e3b:bff:fea5:1, Dst: ff02::1:2
> User Datagram Protocol, Src Port: 546, Dst Port: 547
v DHCPv6
    Message type: Request (3)
    Transaction ID: 0x29b145
  > Client Identifier
  > Server Identifier
  v Identity Association for Non-temporary Address
      Option: Identity Association for Non-temporary Address (3)
      Length: 40
      IAID: 00000041
      T1: 918
      T2: 1458
    v IA Address
        Option: IA Address (5)
        Length: 24
        IPv6 address: 2000::2
        Preferred lifetime: 1800
        Valid lifetime: 3600
  v Option Request
      Option: Option Request (6)
      Length: 8
      Requested Option code: Vendor-specific Information (17)
      Requested Option code: DNS recursive name server (23)
      Requested Option code: Simple Network Time Protocol Server (31)
      Requested Option code: CAPWAP Access Controllers (52)
  > Elapsed time
```

图 9-6　DHCPv6 首次分配地址时第 3 次交互对应的 Request 报文

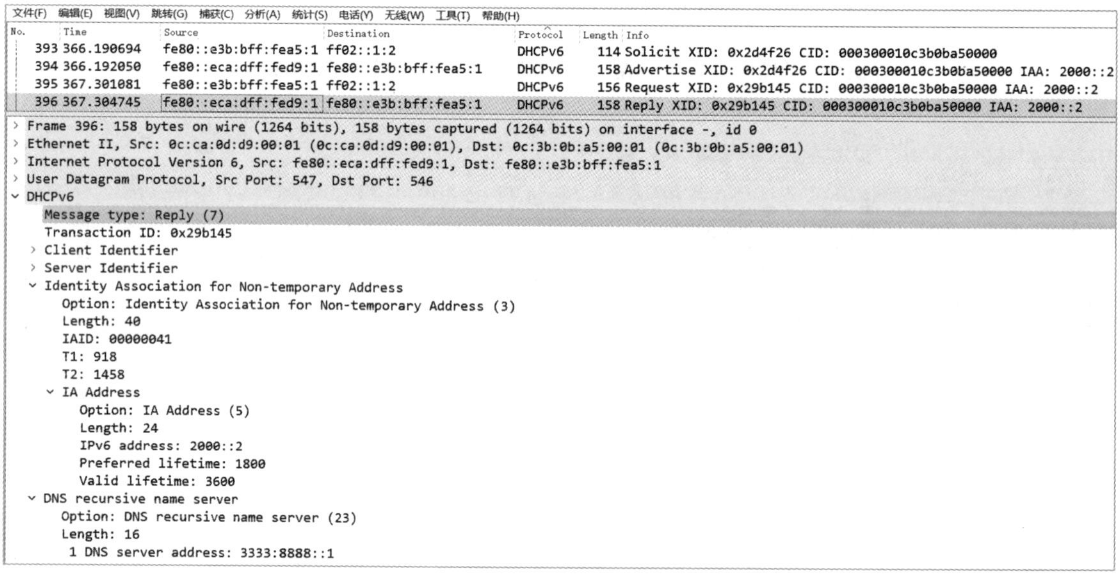

```
文件(F)  编辑(E)  视图(V)  跳转(G)  捕获(C)  分析(A)  统计(S)  电话(Y)  无线(W)  工具(T)  帮助(H)
No.    Time          Source                    Destination            Protocol  Length  Info
 393 366.190694   fe80::e3b:bff:fea5:1      ff02::1:2              DHCPv6    114 Solicit  XID: 0x2d4f26 CID: 000300010c3b0ba50000
 394 366.192050   fe80::eca:dff:fed9:1      fe80::e3b:bff:fea5:1   DHCPv6    158 Advertise XID: 0x2d4f26 CID: 000300010c3b0ba50000 IAA: 2000::2
 395 367.301081   fe80::e3b:bff:fea5:1      ff02::1:2              DHCPv6    156 Request  XID: 0x29b145 CID: 000300010c3b0ba50000 IAA: 2000::2
 396 367.304745   fe80::eca:dff:fed9:1      fe80::e3b:bff:fea5:1   DHCPv6    158 Reply    XID: 0x29b145 CID: 000300010c3b0ba50000 IAA: 2000::2

> Frame 396: 158 bytes on wire (1264 bits), 158 bytes captured (1264 bits) on interface -, id 0
> Ethernet II, Src: 0c:ca:0d:d9:00:01 (0c:ca:0d:d9:00:01), Dst: 0c:3b:0b:a5:00:01 (0c:3b:0b:a5:00:01)
> Internet Protocol Version 6, Src: fe80::eca:dff:fed9:1, Dst: fe80::e3b:bff:fea5:1
> User Datagram Protocol, Src Port: 547, Dst Port: 546
v DHCPv6
    Message type: Reply (7)
    Transaction ID: 0x29b145
  > Client Identifier
  > Server Identifier
  v Identity Association for Non-temporary Address
      Option: Identity Association for Non-temporary Address (3)
      Length: 40
      IAID: 00000041
      T1: 918
      T2: 1458
    v IA Address
        Option: IA Address (5)
        Length: 24
        IPv6 address: 2000::2
        Preferred lifetime: 1800
        Valid lifetime: 3600
  v DNS recursive name server
      Option: DNS recursive name server (23)
      Length: 16
      1 DNS server address: 3333:8888::1
```

图 9-7　DHCPv6 首次分配地址时第 4 次交互对应的 Reply 报文

2）DHCPv6 地址更新时的两次交互过程

当接口执行命令"dhcpv6 client renew"进行有状态 IP 地址更新时，DHCPv6 客户端与服务器需要经历两次交互过程，这两次交互对应的报文分别是 Renew 报文（见图 9-8）和 Reply 报文（见图 9-9）。

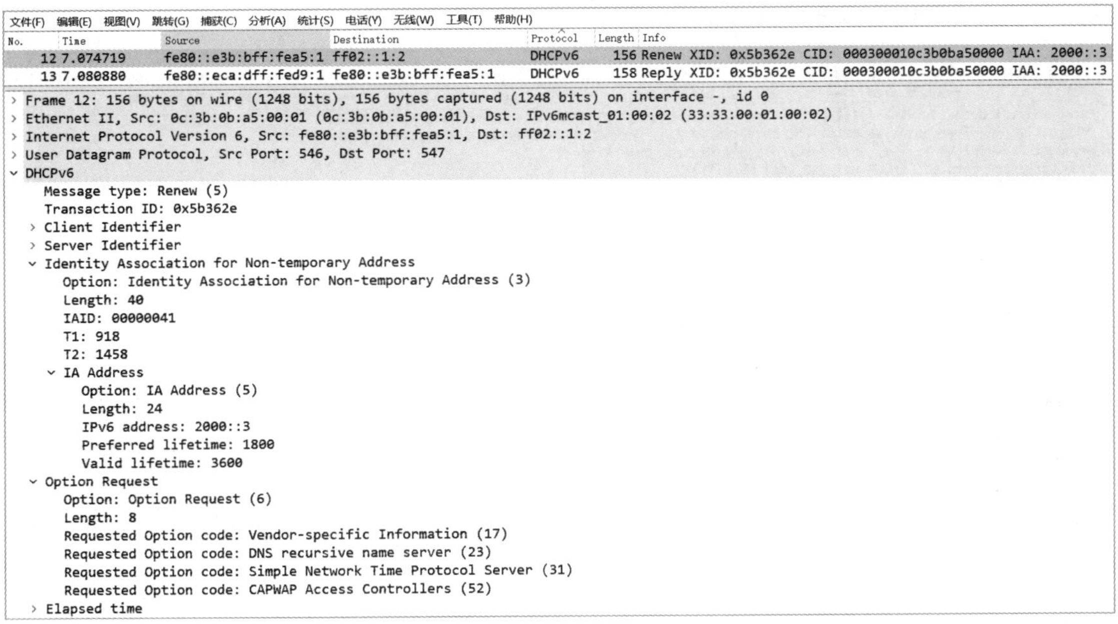

图 9-8 DHCPv6 地址更新时第 1 次交互对应的 Renew 报文

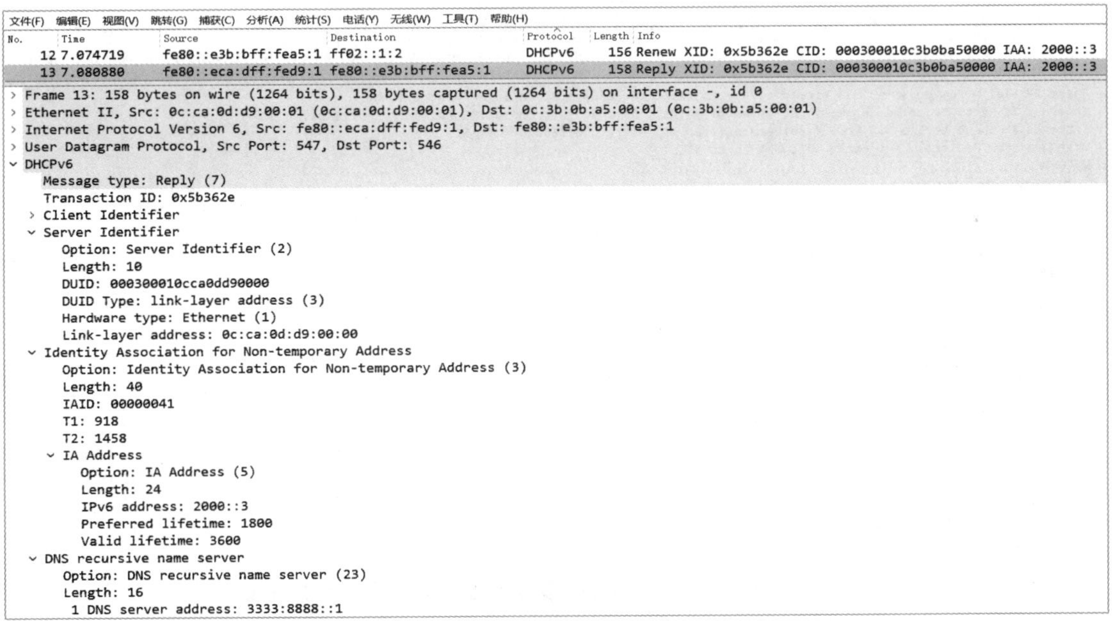

图 9-9 DHCPv6 地址更新时第 2 次交互对应的 Reply 报文

3）DHCPv6 地址释放时的两次交互过程

当接口执行命令"undo ipv6 address auto dhcp"释放 IP 地址时，DHCPv6 客户端与服务器需要经历两次交互过程，这两次交互对应的报文分别是 Release 报文（见图 9-10）和 Reply 报文（见图 9-11）。

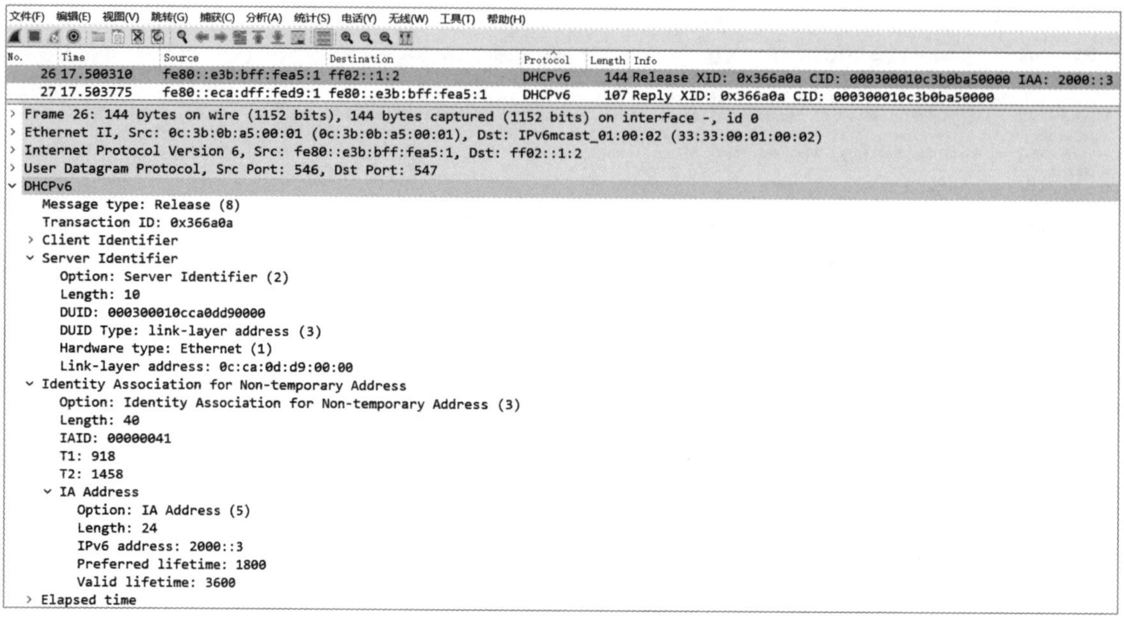

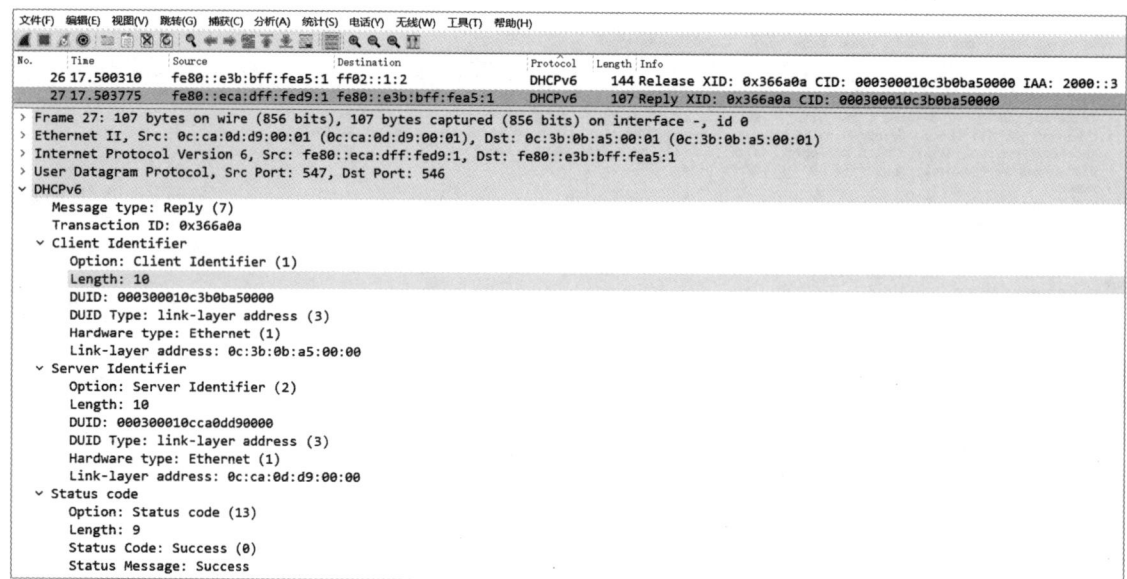

图 9-10　DHCPv6 地址释放时第 1 次交互对应的 Release 报文

图 9-11　DHCPv6 地址释放时第 2 次交互对应的 Reply 报文

第 10 章
基于路由器的 CBAC 防火墙

10.1 组网需求

通过 EIGRP（Enhanced Interior Gateway Routing Protocol，增加型内部网关路由协议），可实现全网在网络层互通。网络拓扑如图 10-1 所示。配置路由器 R2 为基于 IOS 的防火墙，开启 CBAC 功能。其中，接口 Fa0/0 的安全等级较低，为不可信任接口；接口 Fa0/1 的安全等级较高，为可信任接口。不可信任接口默认拒绝所有流量，安全等级高的区域可以正常访问安全等级低的区域网络的任何资源。

一旦流量被接口 Fa0/0 拒绝，R2 便向服务器发送系统日志消息；一旦 R2 监控到"可信任"网络对"不可信任"网络产生 TCP 连接，便立即向服务器发送系统日志消息。

在路由器 R1 上可以通过 telnet 202.10.12.3 远程登录路由器 R3，实现远程管理 R3。

为了防止拒绝服务攻击，在路由器 R2 上还需要实现下列功能。

（1）半开连接数达到 500 个时，丢弃后续所有的 TCP 连接，直到半开连接数降到 300 个以内；当每分钟的半开连接数超过 200 个时，丢弃后续所有的 TCP 连接，直到每分钟的半开连接数降到 100 个以内。

（2）若某主机的半开连接数达到 30 个，则禁止该主机在 10min 内的所有新连接。

（3）丢弃所有 20s 内没有完成三次握手的 TCP 连接；在发现 FIN 报文后，在 4s 内丢弃该连接；TCP 会话空闲达到 30min，则丢弃该连接；UDP 会话空闲 15s，则丢弃该会话；所有数据包的分片数量不得超过 10 个。

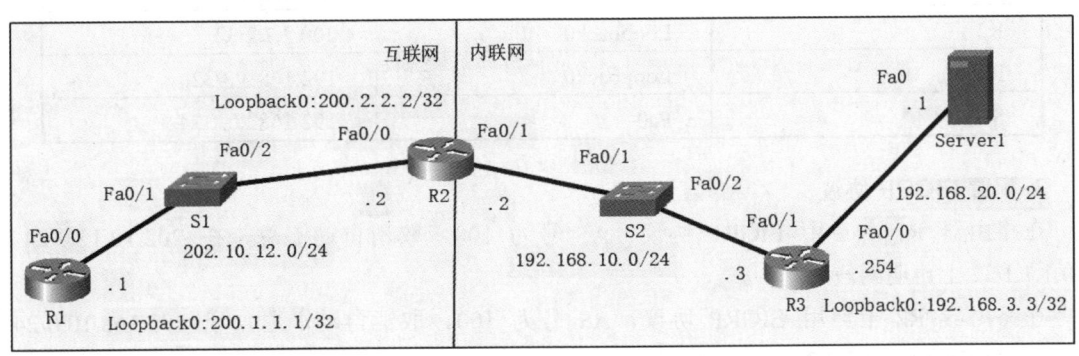

图 10-1 网络拓扑

10.2 仿真实验

10.2.1 仿真环境

仿真环境所需要的硬件、软件建议至少满足以下要求。

- ➢ 物理机的处理器主频为 1.6GHz；
- ➢ 内存 RAM 大小为 32.0GB；
- ➢ 物理机的操作系统为 Windows 10；
- ➢ 虚拟机为 VMware Workstation 16 Pro；
- ➢ 网络模拟器选用 Cisco Packet Tracer，版本 8.0。

10.2.2 实验设计

根据网络拓扑（见图 10-1）完成网络设备的选型：3 台路由器选用 Cisco 2811 系列路由器，两台交换机选用 Cisco WS-C2950-24 系列二层交换机，服务器选用 PT（Packet Tracer）通用型服务器 Server-PT。根据前面的组网需求，本实验设计如下。

1. VLAN 设计

两台交换机无须新建 VLAN，所有接口均属于 VLAN1。

2. IP 地址规划

网络中有 3 个物理子网，网络地址分别为 202.10.12.0/24、192.168.10.0/24、192.168.20.0/24。此外，还有 3 个环回地址。各设备接口的 IP 地址/掩码如表 10-1 所示。

表 10-1 各设备接口的 IP 地址/掩码

设备	接口	IP 地址/掩码
R1	Fa0/0	202.10.12.1/24
R2	Fa0/0	202.10.12.2/24
R2	Fa0/1	192.168.10.2/24
R3	Fa0/1	192.168.10.3/24
R3	Fa0/0	192.168.20.254/24
R1	Loopback0	200.1.1.1/32
R2	Loopback0	200.2.2.2/32
R3	Loopback0	192.168.3.3/32
Server1	Fa0	192.168.20.1/24

3. 配置 EIGRP 协议

在路由器 R1 上启用 EIGRP 协议，AS 号为 100，取消自动汇总，在 202.10.12.0/24 和 200.1.1.1/32 上声明网段。

在路由器 R2 上启用 EIGRP 协议，AS 号为 100，取消自动汇总，在 192.168.10.0/24、202.10.12.0/24 和 200.2.2.2/32 上声明网段。

在路由器 R3 上启用 EIGRP 协议，AS 号为 100，取消自动汇总，在 192.168.10.0/24、

192.168.20.0/24 和 192.168.3.3/32 上声明网段。

4. 防火墙设计

将路由器 R2 配置成 IOS 防火墙，首先在 R2 上开启 CBAC 功能，然后将 Fa0/0 设置为不信任接口，将 Fa0/1 设置为信任接口。当流量被接口 Fa0/0 拒绝时，防火墙向服务器发送系统日志消息。同时，在防火墙上配置允许路由器 R1 远程登录路由器 R3，实现远程管理。在此基础上，为了有效应对拒绝服务攻击，在路由器 R2 上对 TCP 连接数量、同步超时、会话超时、IP 分片数量和 UDP 超时设置限制，以最大可能地避免资源耗尽。

10.2.3 操作步骤

下面给出本次实验的具体操作，一共包括 10 个步骤。

1. 按照图 10-1 所示的网络拓扑，在 PT 中新建一个空白的工程

3 台路由器选用 Cisco 2811 系列路由器，两台交换机选用 Cisco WS-C2950-24 系列二层交换机，服务器选用 PT 通用型服务器 Server-PT。

2. 配置路由器的 IP 地址

1）配置路由器 R1 的 IP 地址

```
Router(config)#hostname cxyR1
cxyR1(config)#interface fa0/0
cxyR1(config-if)#ip address 202.10.12.1 255.255.255.0
cxyR1(config-if)#no shutdown
cxyR1(config-if)#exit
cxyR1(config)#interface loopback0
cxyR1(config-if)#ip address 200.1.1.1 255.255.255.255
cxyR1(config-if)#exit
```

2）配置路由器 R2 的 IP 地址

```
Router(config)#hostname cxyR2
cxyR2(config)#interface fa0/0
cxyR2(config-if)#ip address 202.10.12.2 255.255.255.0
cxyR2(config-if)#no shutdown
cxyR2(config-if)#exit
cxyR2(config)#interface fa0/1
cxyR2(config-if)#ip address 192.168.10.2 255.255.255.0
cxyR2(config-if)#no shutdown
cxyR2(config-if)#exit
cxyR2(config)#interface loopback0
cxyR2(config-if)#ip address 200.2.2.2 255.255.255.255
cxyR2(config-if)#exit
```

3）配置路由器 R3 的 IP 地址

```
Router(config)#hostname cxyR3
cxyR3(config)#interface fa0/0
cxyR3(config-if)#ip address 192.168.20.254 255.255.255.0
cxyR3(config-if)#no shutdown
cxyR3(config-if)#exit
cxyR3(config)#interface fa0/1
cxyR3(config-if)#ip address 192.168.10.3 255.255.255.0
cxyR3(config-if)#no shutdown
cxyR3(config-if)#exit
cxyR3(config)#interface loopback0
cxyR3(config-if)#ip address 192.168.3.3 255.255.255.255
cxyR3(config-if)#exit
```

4）测试直连网段的连通性

```
cxyR1#ping 202.10.12.2
Type escape sequence to abort
Sending 5, 100-byte ICMP Echos to 202.10.12.2, timeout is 2 seconds:
.!!!!
Success rate is 80 percent (4/5), round-trip min/avg/max = 1/2/4 ms
cxyR2#ping 192.168.10.3
Type escape sequence to abort.
Sending 5, 100-byte ICMP Echos to 192.168.10.3, timeout is 2 seconds:
.!!!!
Success rate is 80 percent (4/5), round-trip min/avg/max = 1/1/4 ms
cxyR3#ping 192.168.20.1
Type escape sequence to abort.
Sending 5, 100-byte ICMP Echos to 192.168.20.1, timeout is 2 seconds:
.!!!!
Success rate is 80 percent (4/5), round-trip min/avg/max = 1/2/4 ms
```

3. 配置 EIGRP 协议，实现全网在网络层互通

1）配置路由器 R1 的 EIGRP 协议

```
cxyR1(config)#router eigrp 100                              //启用 EIGRP 协议，AS 号为 100
cxyR1(config-router)#no auto-summary
cxyR1(config-router)#network 202.10.12.0  0.0.0.255         //宣告网段
cxyR1(config-router)#network 200.1.1.1  0.0.0.0
cxyR1(config-router)#exit
```

2）配置路由器 R2 的 EIGRP 协议

```
cxyR2(config)#router eigrp 100
cxyR2(config-router)#no auto-summary
cxyR2(config-router)#network 202.10.12.0 0.0.0.255
cxyR2(config-router)#network 192.168.10.0 0.0.0.255
cxyR2(config-router)#network 200.2.2.2 0.0.0.0
cxyR2(config-router)#exit
```

3）配置路由器 R3 的 EIGRP 协议

```
cxyR3(config)#router eigrp 100
cxyR3(config-router)#no auto-summary
cxyR3(config-router)#network 192.168.10.0 0.0.0.255
cxyR3(config-router)#network 192.168.3.3 0.0.0.0
cxyR3(config-router)#network 192.168.20.254 0.0.0.255
cxyR3(config-router)#exit
```

4）查看路由器 R1 的路由表

```
cxyR1#show ip route
Codes: C - connected, S - static, R - RIP, M - mobile, B - BGP
D - EIGRP, EX - EIGRP external, O - OSPF, IA - OSPF inter area
N1 - OSPF NSSA external type 1, N2 - OSPF NSSA external type 2
E1 - OSPF external type 1, E2 - OSPF external type 2
i - IS-IS, su - IS-IS summary, L1 - IS-IS level-1, L2 - IS-IS level-2
ia - IS-IS inter area, * - candidate default, U - per-user static route
o - ODR, P - periodic downloaded static route
Gateway of last resort is not set
C  202.10.12.0/24 is directly connected, FastEthernet0/0
D  192.168.10.0/24 [90/30720] via 202.10.12.2, 00:01:59, FastEthernet0/0
   192.168.0.0/32 is subnetted, 3 subnets
C  200.1.1.1 is directly connected, Loopback0
D  200.2.2.2 [90/156160] via 202.10.12.2, 00:01:56, FastEthernet0/0
D  192.168.3.3 [90/158720] via 202.10.12.2, 00:01:29, FastEthernet0/0
D  192.168.20.0/24 [90/33280] via 202.10.12.2, 00:01:15, FastEthernet0/0
```

4. 配置路由器 R2 为 IOS 防火墙，开启 CBAC 功能，设置 Fa0/0 为不信任接口、Fa0/1 为信任接口

```
cxyR2(config)#ip access-list extended cxyCBAC_ACL
cxyR2(config-ext-nacl)#permit eigrp host 202.10.12.1 host 224.0.0.10   //224.0.0.10 为 EIGRP 组
                                                                       //播组地址
```

```
cxyR2(config-ext-nacl)#permit eigrp host 202.10.12.1 host 202.10.12.2
cxyR2(config-ext-nacl)#exit
cxyR2(config)#interface fa0/0   //fa0/0 为不信任接口，拒绝除 EIGRP 外的所有流量
cxyR2(config-if)#ip access-group cxyCBAC_ACL in
cxyR2(config-if)#exit
//信任端可以正常访问非信任网络的所有资源（TCP、UDP、ICMP）
cxyR2(config)#ip inspect name cxyCBAC tcp          //定义名为 cxyCBAC 的监控策略
cxyR2(config)#ip inspect name cxyCBAC udp
cxyR2(config)#ip inspect name cxyCBAC icmp
cxyR2(config)#interface fa0/0
cxyR2(config-if)#ip inspect cxyCBAC out            //在接口上应用监控策略
cxyR2(config-if)#exit
```

5. 查看 CBAC

```
cxyR2#show ip inspect all
Session audit trail is disabled
Session alert is enabled
one-minute (sampling period) thresholds are [unlimited : unlimited] connections
max-incomplete sessions thresholds are [unlimited : unlimited]
max-incomplete tcp connections per host is unlimited. Block-time 0 minute.
tcp synwait-time is 30 sec -- tcp finwait-time is 5 sec
tcp idle-time is 3600 sec -- udp idle-time is 30 sec
dns-timeout is 5 sec
Inspection Rule Configuration
Inspection name cxyCBAC
tcp alert is on audit-trail is off timeout 3600
udp alert is on audit-trail is off timeout 30
icmp alert is on audit-trail is off timeout 10

Interface Configuration
Interface FastEthernet0/0
Inbound inspection rule is not set
Outgoing inspection rule is cxyCBAC
tcp alert is on audit-trail is off timeout 3600
udp alert is on audit-trail is off timeout 30
icmp alert is on audit-trail is off timeout 10
Inbound access list is cxyCBAC_ACL
Outgoing access list is not set
```

第 10 章 基于路由器的 CBAC 防火墙

6. 测试

```
cxyR1#ping 192.168.3.3                           //数据包被拒绝
Type escape sequence to abort.
Sending 5, 100-byte ICMP Echos to 192.168.3.3, timeout is 2 seconds:
U.U.U
Success rate is 0 percent (0/5)
cxyR3#ping 200.1.1.1
Type escape sequence to abort.
Sending 5, 100-byte ICMP Echos to 200.1.1.1, timeout is 2 seconds:
!!!!!
Success rate is 100 percent (5/5), round-trip min/avg/max = 1/4/8 ms
```

若路由器 R2 未配置命令"ip inspect name cxyCBAC icmp",上述命令是否可 ping 通?即 "no ip inspect name cxyCBAC icmp",再 ping 一下试试看。

```
cxyR2#show ip inspect sessions detail
Established Sessions
Session 64462AA8 (192.168.10.3:8)=>(200.1.1.1:0) icmp SIS_OPEN
Created 00:00:10, Last heard 00:00:10
  ECHO request
Bytes sent (initiator:responder) [360:360]
In SID 200.1.1.1[0:0]=>192.168.10.3[0:0] on ACL cxyCBAC_ACL (5 matches)
In SID 0.0.0.0[0:0]=>192.168.10.3[3:3] on ACL cxyCBAC_ACL
In SID 0.0.0.0[0:0]=>192.168.10.3[11:11] on ACL cxyCBAC_ACL
```

7. 当流量被接口 Fa0/0 拒绝时,路由器 R2 向服务器发送系统日志消息

1)开启日志功能,并将日志消息发送至指定服务器

```
cxyR2(config)#logging buffered
cxyR2(config)#logging host 192.168.20.1         //192.168.20.1 为服务器
cxyR2(config)#ip access-list extended cxyCBAC_ACL
cxyR2(config-ext-nacl)#deny ip any any log
cxyR2(config-ext-nacl)#exit
```

2)测试

(1) 在路由器 R1 上 ping192.168.3.3,以触发 R2 产生日志。

```
cxyR1#ping 192.168.3.3
Type escape sequence to abort.
Sending 5, 100-byte ICMP Echos to 192.168.3.3, timeout is 2 seconds:
U.U.U
Success rate is 0 percent (0/5)
```

（2）在路由器 R2 上查看日志。

cxyR2#show logging

Syslog logging: enabled (11 messages dropped, 0 messages rate-limited, 0 flushes, 0 overruns, xml disabled, filtering disabled)

Console logging: level debugging, 61 messages logged, xml disabled, filtering disabled

Monitor logging: level debugging, 0 messages logged, xml disabled, filtering disabled

Buffer logging: level debugging, 3 messages logged, xml disabled, filtering disabled

Logging Exception size (4096 bytes)

Count and timestamp logging messages: disabled

No active filter modules.

Trap logging: level informational, 64 message lines logged

Logging to 192.168.20.1(global) (udp port 514, audit disabled, link up), 3 message lines logged, xml disabled, filtering disabled

Log Buffer (4096 bytes):

*Jul 20 11:00:30.390: %SEC-6-IPACCESSLOGDP: list cxyCBAC_ACL denied icmp 202.10.12.1 -> 192.168.3.3 (8/0), 1 packet

*Jul 20 11:00:31.390: %SYS-6-LOGGINGHOST_STARTSTOP: Logging to host 192.168.20.1 started - CLI initiated

*Jul 20 11:00:33.866: %SYS-5-CONFIG_I: Configured from console by console

8. 一旦路由器 R2 监控到信任网络对非信任网络产生 TCP 连接，便立即向服务器发送系统日志消息

1）在 R2 上配置 CBAC

cxyR2(config)#ip inspect name cxyCBAC tcp audit-trail on

2）在路由器 R3 上执行 telnet 200.1.1.1，以触发 R2 产生日志

cxyR3#telnet 200.1.1.1

Trying 200.1.1.1 ... Open

cxyR1>

cxyR1>exit

[Connection to 200.1.1.1 closed by foreign host]

cxyR3#

3）查看路由器 R2 的日志信息

cxyR2#show logging

Syslog logging: enabled (11 messages dropped, 0 messages rate-limited, 0 flushes, 0 overruns, xml disabled, filtering disabled)

Console logging: level debugging, 64 messages logged, xml disabled, filtering disabled

Monitor logging: level debugging, 0 messages logged, xml disabled, filtering disabled

Buffer logging: level debugging, 6 messages logged, xml disabled, filtering disabled

Logging Exception size (4096 bytes)

Count and timestamp logging messages: disabled

No active filter modules.

Trap logging: level informational, 67 message lines logged

Logging to 192.168.20.1(global) (udp port 514, audit disabled, link up), 6 message lines logged, xml disabled, filtering disabled

Log Buffer (4096 bytes):

*Jul 20 11:00:30.390: %SEC-6-IPACCESSLOGDP: list cxyCBAC_ACL denied icmp 202.10.12.1 -> 192.168.3.3 (8/0), 1 packet

*Jul 20 11:00:31.390: %SYS-6-LOGGINGHOST_STARTSTOP: Logging to host 192.168.20.1 started - CLI initiated

*Jul 20 11:00:33.866: %SYS-5-CONFIG_I: Configured from console by console

*Jul 20 11:04:37.494: %FW-6-SESS_AUDIT_TRAIL_START: Start tcp session: initiator (192.168.10.3:11002) -- responder (200.1.1.1:23)

*Jul 20 11:04:44.414: %SYS-5-CONFIG_I: Configured from console by console

*Jul 20 11:04:44.802: %FW-6-SESS_AUDIT_TRAIL: Stop tcp session: initiator (192.168.10.3:11002) sent 40 bytes -- responder (200.1.1.1:23) sent 54 bytes

9. 设置路由器 R1 能够远程登录路由器 R3，实现远程管理

1）允许在路由器 R1 上通过执行 telnet 202.10.12.3，远程登录路由器 R3

cxyR2(config)#ip nat inside source static tcp 192.168.10.3 23 202.10.12.3 23

cxyR2(config)#interface fastEthernet 0/0

cxyR2(config-if)#ip nat outside

cxyR2(config-if)#exit

cxyR2(config)#interface fastEthernet 0/1

cxyR2(config-if)#ip nat inside

cxyR2(config-if)#exit

cxyR2(config)#ip access-list extended cxyCBAC_ACL

cxyR2(config-ext-nacl)#25 permit tcp host 202.10.12.1 host 192.168.10.3 eq 23

cxyR2(config-ext-nacl)#exit

2）从 R1 上测试远程登录 R3

cxyR1#telnet 202.10.12.3

Trying 202.10.12.3 ... Open

cxyR3>

cxyR3>exit

[Connection to 202.10.12.3 closed by foreign host]
cxyR1#

10. 配置 TCP、UDP、IP 的限制

1）当半开连接数超过 1000 时，丢弃后续所有的 TCP 连接，直到半开连接数小于 500

cxyR2(config)#ip inspect max-incomplete high 1000
cxyR2(config)#ip inspect max-incomplete low 500
// 查看最大半开连接数的阈值
cxyR2#show ip inspect config
Session audit trail is disabled
Session alert is enabled
one-minute (sampling period) thresholds are [100 : 200] connections
max-incomplete sessions thresholds are [500 : 1000]
max-incomplete tcp connections per host is unlimited. Block-time 0 minute.
tcp synwait-time is 20 sec -- tcp finwait-time is 4 sec
tcp idle-time is 3600 sec -- udp idle-time is 30 sec
dns-timeout is 5 sec
Inspection Rule Configuration
Inspection name cxyCBAC
tcp alert is on audit-trail is off timeout 1800
udp alert is on audit-trail is off timeout 15
icmp alert is on audit-trail is off timeout 10

2）当每分钟的半开连接数超过 400 时，丢弃后续所有的 TCP 连接，直到该数值小于 200

cxyR2(config)#ip inspect one-minute high 200
cxyR2(config)#ip inspect one-minute low 100
// 查看每分钟半开连接数的阈值
cxyR2#show ip inspect config
Session audit trail is disabled
Session alert is enabled
one-minute (sampling period) thresholds are [200 : 400] connections
max-incomplete sessions thresholds are [500 : 1000]
max-incomplete tcp connections per host is unlimited. Block-time 0 minute.
tcp synwait-time is 20 sec -- tcp finwait-time is 4 sec
tcp idle-time is 3600 sec -- udp idle-time is 30 sec
dns-timeout is 5 sec
Inspection Rule Configuration
Inspection name cxyCBAC
tcp alert is on audit-trail is off timeout 1800

udp alert is on audit-trail is off timeout 15

icmp alert is on audit-trail is off timeout 10

3）丢弃所有 15s 内没有完成三次握手的 TCP 连接；在收到 FIN 报文后，5s 内丢弃该连接

cxyR2(config)#ip inspect tcp synwait-time 15

cxyR2(config)#ip inspect tcp finwait-time 5

// 查看建立连接和释放连接的等待时间

cxyR2#show ip inspect config

Session audit trail is disabled

Session alert is enabled

one-minute (sampling period) thresholds are [200 : 400] connections

max-incomplete sessions thresholds are [500 : 1000]

max-incomplete tcp connections per host is unlimited. Block-time 0 minute.

tcp synwait-time is 15 sec -- tcp finwait-time is 5 sec

tcp idle-time is 3600 sec -- udp idle-time is 30 sec

dns-timeout is 5 sec

Inspection Rule Configuration

Inspection name cxyCBAC

tcp alert is on audit-trail is off timeout 1800

udp alert is on audit-trail is off timeout 15

icmp alert is on audit-trail is off timeout 10

4）当 TCP 会话空闲达到 1200s 时，丢弃该连接；UDP 会话空闲超时为 20s

cxyR2(config)#ip inspect tcp idle-time 1200

cxyR2(config)#ip inspect udp idle-time 20

// 查看 TCP 会话和 UDP 会话的空闲时长

cxyR2#show ip inspect config

Session audit trail is disabled

Session alert is enabled

one-minute (sampling period) thresholds are [200 : 400] connections

max-incomplete sessions thresholds are [500 : 1000]

max-incomplete tcp connections per host is unlimited. Block-time 0 minute.

tcp synwait-time is 15 sec -- tcp finwait-time is 5 sec

tcp idle-time is 1200 sec -- udp idle-time is 20 sec

dns-timeout is 5 sec

Inspection Rule Configuration

Inspection name cxyCBAC

tcp alert is on audit-trail is off timeout 1800

udp alert is on audit-trail is off timeout 15

icmp alert is on audit-trail is off timeout 10

5）若某主机的半开连接数达到 30，则禁止该主机在 15min 内发起新连接

cxyR2(config)#ip inspect tcp max-incomplete host 30 block-time 15
// 查看主机半开连接数最大值和拒绝连接时长
cxyR2#show ip inspect config
Session audit trail is disabled
Session alert is enabled
one-minute (sampling period) thresholds are [200 : 400] connections
max-incomplete sessions thresholds are [500 : 1000]
max-incomplete tcp connections per host is 30. Block-time 15 minute.
tcp synwait-time is 15 sec -- tcp finwait-time is 5 sec
tcp idle-time is 1200 sec -- udp idle-time is 20 sec
dns-timeout is 5 sec
Inspection Rule Configuration
Inspection name cxyCBAC
tcp alert is on audit-trail is off timeout 1800
udp alert is on audit-trail is off timeout 15
icmp alert is on audit-trail is off timeout 10

6）所有分组的分片数量不得超过 10 个

cxyR2(config)#ip inspect name cxyCBAC fragment maximum 10
//查看允许分片数量的最大值
cxyR2#show ip inspect config
Session audit trail is disabled
Session alert is enabled
one-minute (sampling period) thresholds are [200 : 400] connections
max-incomplete sessions thresholds are [500 : 1000]
max-incomplete tcp connections per host is 30. Block-time 15 minute.
tcp synwait-time is 15 sec -- tcp finwait-time is 5 sec
tcp idle-time is 1200 sec -- udp idle-time is 20 sec
dns-timeout is 5 sec
Inspection Rule Configuration
Inspection name cxyCBAC
tcp alert is on audit-trail is off timeout 1800
udp alert is on audit-trail is off timeout 15
icmp alert is on audit-trail is off timeout 10
fragment Maximum 10 In Use 0 alert is on audit-trail is off timeout 1

第 11 章
基于路由器的 IPsec VPN

11.1 组网需求

网络拓扑如图 11-1 所示,公司总部的私有网络连接在路由器 R1 的 Fa0/1 接口上,公司分部的私有网络连接在路由器 R2 上。互联网是由 R1、R2、R4 和交换机 SW 组成的网络模拟的。

现要求在路由器 R1 和 R2 间配置站点到站点的 IPsec VPN,加密算法选择 3DES,Hash 函数使用 SHA,认证使用预共享密钥并在两个路由器上设置相同的共享密钥。在 R1 和 R2 上要求对 192.168.3.0/24 和 192.168.2.0/24 两个网段之间的流量进行加密。VPN 建立成功后,公司分部网段内的用户可以访问公司总部网络的内部资源。

在此基础上,要求在路由器 R1 的 Fa0/0 接口上应用访问控制列表,只允许 ICMP 和 Telnet 报文通过,同时不能影响 VPN 流量。

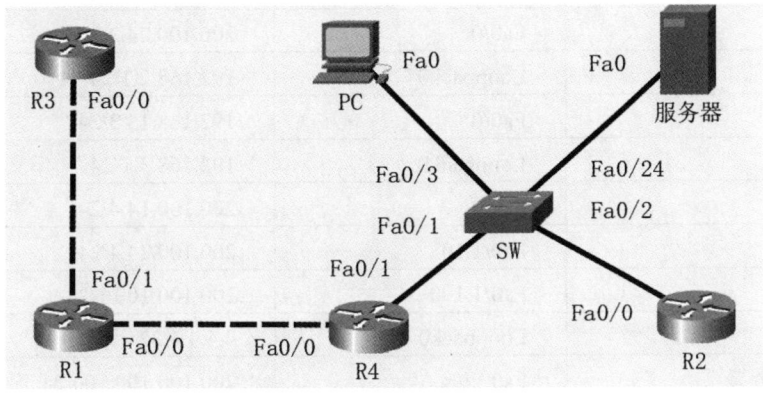

图 11-1 网络拓扑

11.2 仿真实验

11.2.1 仿真环境

仿真环境所需要的硬件、软件建议至少满足以下要求。

- ➢ 物理机的处理器主频为 1.6GHz;
- ➢ 内存 RAM 大小为 32.0GB;
- ➢ 物理机的操作系统为 Windows 10;

> 网络模拟器选用 Cisco Packet Tracer,版本 8.0。

11.2.2 实验设计

根据网络拓扑(见图 11-1)完成网络设备的选型:4 台路由器选用 Cisco 2811 系列路由器,1 台交换机选用 Cisco WS-C2950-24 系列二层交换机,服务器选用通用型服务器 Server-PT。根据前面的组网需求,本实验设计如下。

1. VLAN 设计

在交换机上新建两个 VLAN:VLAN10 和 VLAN100,将接口 Fa0/2 和 Fa0/3 划分到 VLAN10 中,把接口 Fa0/24 划分到 VLAN100 中。将接口 Fa0/1 设置为 trunk 模式并允许所有 VLAN 通过。

2. IP 地址规划

网络中有 4 个物理子网,网络地址分别为 200.100.14.0/24、192.168.13.0/24、200.100.24.0/24 和 200.100.100.0/24。此外,还有 4 个环回地址。各设备接口的 IP 地址/掩码如表 11-1 所示。

表 11-1 各设备接口的 IP 地址/掩码

设备	接口	IP 地址/掩码
R1	Fa0/0	200.100.14.1/24
R1	Fa0/1	192.168.13.1/24
R1	Loopback0	192.168.1.1/24
R2	Fa0/0	200.100.24.2/24
R2	Loopback0	192.168.2.1/24
R3	Fa0/0	192.168.13.3/24
R3	Loopback0	192.168.3.1/24
R4	Fa0/0	200.100.14.4/24
R4	Fa0/1.10	200.100.24.4/24
R4	Fa0/1.100	200.100.100.1/24
R4	Loopback0	4.4.4.4/32
服务器	Fa0	200.100.100.100/24

3. 路由设计

在公网上配置 EIGRP 协议,使得公网所有 IP 地址在网络层互通:在 R1、R2 和 R4 上分别启用 EIGRP 协议,AS 号均为 100,取消自动汇总。R1 在 200.100.14.0/24 上声明网段,R2 在 200.100.24.0/24 上声明网段,R4 在 200.100.14.0/24、200.100.24.0/24 和 200.100.100.0/24 上声明网段。

在内网上配置 RIP(Routing Information Protocol,路由信息协议),使得内网所有 IP 地址在网络层互通:在 R1 和 R3 上分别启用 RIPv2 协议,取消自动汇总。R1 在 192.168.13.0/24 和 192.168.1.0/24 上声明网段,R3 在 192.168.13.0/24 和 192.168.3.0/24 上声明网段。

4. IPsec VPN 设计

在路由器 R1 和 R2 间配置站点到站点的 IPsec VPN，具体包括在路由器 R1 和 R2 上分别开启并配置 ISAKMP（Internet Security Association Key Management Protocol，互联网安全关联密钥管理协议），加密算法选择 3DES，Hash 函数使用 SHA，认证使用预共享密钥并在两台路由器上设置相同的共享密钥。在 R1 和 R2 上定义需要加密的数据流，要求对 192.168.3.0/24 和 192.168.2.0/24 两个网段之间的流量进行加密。为此，在两台路由器上定义两条互为镜像的 ACL 规则。

11.2.3 操作步骤

下面给出本次实验的具体操作，一共包括 9 个步骤。

1. 按照图 11-1 所示的网络拓扑，在 PT 中新建一个空白的工程

4 台路由器选用 Cisco 2811 系列路由器，1 台交换机选用 Cisco WS-C2950-24 系列二层交换机，服务器选用 PT 通用型服务器 Server-PT。

2. 配置交换机，划分 VLAN 并把接口划分到相应的 VLAN 中

```
SW(config)#vlan10
SW(config-vlan)#exit
SW(config)#vlan100
SW(config-vlan)#exit
SW(config)#interface fa0/1     //交换机的该接口与 R4 的 Fa0/1 相连
SW(config-if)#switchport mode trunk
SW(config-if)#exit
SW(config)#interface fa0/2     //交换机的该接口与 R2 的 Fa0/0 相连
SW(config-if)#switchport mode access
SW(config-if)#switchport access vlan10
SW(config)#interface fa0/3     //交换机的该接口与 PC 的 Fa0 相连
SW(config-if)#switchport mode access
SW(config-if)#switchport access vlan10
SW(config-if)#exit
SW(config)#interface fa0/24    //交换机的该接口与服务器的 Fa0 相连
SW(config-if)#switchport mode access
SW(config-if)#switchport access vlan100
SW(config-if)#exit
```

3. 配置路由器的 IP 地址

1）配置路由器 R1 的 IP 地址

```
R1(config)#interface fa0/0
R1(config-if)#ip address 200.100.14.1 255.255.255.0
R1(config-if)#no shutdown
```

R1(config-if)#exit
R1(config)#interface fa0/1
R1(config-if)#ip address 192.168.13.1 255.255.255.0
R1(config-if)#no shutdown
R1(config-if)#exit
R1(config)#interface loopback0
R1(config-if)#ip address 192.168.1.1 255.255.255.0
R1(config-if)#exit

2）配置路由器 R2 的 IP 地址

R2(config)#interface fa0/0
R2(config-if)#ip address 200.100.24.2 255.255.255.0
R2(config-if)#no shutdown
R2(config-if)#exit
R2(config)#interface loopback0
R2(config-if)#ip address 192.168.2.1 255.255.255.0
R2(config-if)#exit

3）配置路由器 R3 的 IP 地址

R3(config)#interface fa0/0
R3(config-if)#ip address 192.168.13.3 255.255.255.0
R3(config-if)#no shutdown
R3(config-if)#exit
R3(config)#interface loopback0
R3(config-if)#ip address 192.168.3.1 255.255.255.0
R3(config-if)#exit

4）配置路由器 R4 的 IP 地址

R4(config)#interface fa0/0
R4(config-if)#ip address 200.100.14.4 255.255.255.0
R4(config-if)#no shutdown
R4(config-if)#exit
R4(config)#interface fa0/1
R4(config-if)#no shutdown
R4(config-if)#exit
R4(config)#interface fa0/1.10
R4(config-subif)#encapsulation dot1q 10
R4(config-subif)#ip address 200.100.24.4 255.255.255.0
R4(config-subif)#exit
R4(config)#interface fa0/1.100

R4(config-subif)#encapsulation dot1q 100
R4(config-subif)#ip address 200.100.100.1 255.255.255.0
R4(config-subif)#exit
R4(config)#interface loopback0
R4(config-if)#ip address 4.4.4.4 255.255.255.255
R4(config-if)#exit

5）测试直连网段的连通性

```
R3#ping 192.168.13.1
Type escape sequence to abort.
Sending 5, 100-byte ICMP Echos to 192.168.13.1, timeout is 2 seconds:
.!!!!
Success rate is 80 percent (4/5), round-trip min/avg/max = 1/2/4 ms
R1#ping 200.100.14.4
Type escape sequence to abort.
Sending 5, 100-byte ICMP Echos to 200.100.14.4, timeout is 2 seconds:
.!!!!
Success rate is 80 percent (4/5), round-trip min/avg/max = 1/1/4 ms
R4#ping 200.100.24.2
Type escape sequence to abort.
Sending 5, 100-byte ICMP Echos to 200.100.24.2, timeout is 2 seconds:
.!!!!
Success rate is 80 percent (4/5), round-trip min/avg/max = 1/2/4 ms
```

4. 在公网上配置 EIGRP 协议，使得公网网段在网络层互通

1）在 R1 上配置 EIGRP 协议

```
R1(config)#router eigrp 100
R1(config-router)#no auto-summary
R1(config-router)#network 200.100.14.0 0.0.0.255
R1(config-router)#exit
```

2）在 R2 上配置 EIGRP 协议

```
R2(config)#router eigrp 100
R2(config-router)#no auto-summary
R2(config-router)#network 200.100.24.0 0.0.0.255
R2(config-router)#exit
```

3）在 R4 上配置 EIGRP 协议

```
R4(config)#router eigrp 100
R4(config-router)#no auto-summary
```

R4(config-router)#network 200.100.14.0 0.0.0.255
R4(config-router)#network 200.100.24.0 0.0.0.255
R4(config-router)#network 200.100.100.0 0.0.0.255
R4(config-router)#exit

4）测试 R1 和 R2 的公网地址是否可以互通

R1#ping 200.100.24.2
Type escape sequence to abort.
Sending 5, 100-byte ICMP Echos to 200.100.24.2, timeout is 2 seconds:
!!!!!
Success rate is 100 percent (5/5), round-trip min/avg/max = 1/2/4 ms
R2#ping 200.100.14.1
Type escape sequence to abort.
Sending 5, 100-byte ICMP Echos to 200.100.14.1, timeout is 2 seconds:
!!!!!
Success rate is 100 percent (5/5), round-trip min/avg/max = 1/3/4 ms

5. 在内网上配置 RIP 协议，使得内网网段在网络层互通

1）在 R1 上配置 RIP 协议

R1(config)#router rip
R1(config-router)#version 2
R1(config-router)#no auto-summary
R1(config-router)#network 192.168.13.0
R1(config-router)#network 192.168.1.0
R1(config-router)#exit

2）在 R3 上配置 RIP 协议

R3(config)#router rip
R3(config-router)#version 2
R3(config-router)#no auto-summary
R3(config-router)#network 192.168.13.0
R3(config-router)#network 192.168.3.0
R3(config-router)#exit

6. 在路由器 R1 和 R2 间配置站点到站点的 IPsec VPN

1）在路由器 R1 上配置

（1）开启并配置 ISAKMP 协议。

R1(config)#crypto isakmp enable
R1(config)#crypto isakmp policy 10
R1(config-isakmp)#encryption 3des   // 加密算法选择 3DES

```
R1(config-isakmp)#hash sha                                          // Hash 函数使用 SHA
R1(config-isakmp)#authentication pre-share                          // 认证使用预共享密钥
R1(config-isakmp)#exit
R1(config)#crypto isakmp key CXY123 address 200.100.24.2            // 预共享密钥为 CXY123
```

（2）配置 IPsec 变换集。

```
R1(config)#crypto ipsec transform-set cxy_lab esp-3des esp-sha-hmac
R1(cfg-crypto-trans)#exit
```

（3）定义需要加密的流量。

```
R1(config)#ip access-list extended cxy_vpn
R1(config-ext-nacl)#permit ip 192.168.3.0 0.0.0.255 192.168.2.0 0.0.0.255
R1(config-ext-nacl)#exit
```

（4）定义保密映射。

```
R1(config)#crypto map ipsec_vpn 10 ipsec-isakmp
% NOTE: This new crypto map will remain disabled until a peer and a valid access list have been configured.
R1(config-crypto-map)#set peer 200.100.24.2
R1(config-crypto-map)#set transform-set cxy_lab
R1(config-crypto-map)#match address cxy_vpn
R1(config-crypto-map)#exit
```

（5）在接口上应用保密映射。

```
R1(config)#interface fa0/0
R1(config-if)#crypto map ipsec_vpn
R1(config-if)#exit
```

2）在路由器 R2 上配置

（1）开启并配置 ISAKMP 协议。

```
R2(config)#crypto isakmp enable
R2(config)#crypto isakmp policy 10
R2(config-isakmp)#encryption 3des                                   // 加密算法选择 3DES
R2(config-isakmp)#hash sha                                          // Hash 函数使用 SHA
R2(config-isakmp)#authentication pre-share                          // 认证使用预共享密钥
R2(config-isakmp)#exit
R2(config)#crypto isakmp key CXY123 address 200.100.14.1            // 预共享密钥为 CXY123
```

（2）配置 IPsec 变换集。

```
R2(config)#crypto ipsec transform-set cxy_lab esp-3des esp-sha-hmac
R2(cfg-crypto-trans)#exit
```

（3）定义需要加密的流量。

R2(config)#ip access-list extended cxy_vpn
R2(config-ext-nacl)#permit ip 192.168.2.0 0.0.0.255 192.168.3.0 0.0.0.255
R2(config-ext-nacl)#exit

（4）定义保密映射。

R2(config)#crypto map ipsec_vpn 10 ipsec-isakmp
% NOTE: This new crypto map will remain disabled until a peer and a valid access list have been configured.
R2(config-crypto-map)#set peer 200.100.14.1
R2(config-crypto-map)#set transform-set cxy_lab
R2(config-crypto-map)#match address cxy_vpn
R2(config-crypto-map)#exit

（5）在接口上应用保密映射。

R2(config)#interface fa0/0
R2(config-if)#crypto map ipsec_vpn
R2(config-if)#exit

7. 在 R1 的 Fa0/0 接口上配置 ACL，只允许 ICMP 和 Telnet 访问，同时放行 VPN 报文

R1(config)#ip access-list extended outside
// 允许 ICMP 报文通过
R1(config-ext-nacl)#permit icmp any host 200.100.14.1
// 允许 Telnet 远程登录
R1(config-ext-nacl)#permit tcp any host 200.100.14.1 eq 23
// 允许 EIGRP 报文通过
R1(config-ext-nacl)#permit eigrp host 200.100.14.4 host 200.100.14.1
R1(config-ext-nacl)#permit eigrp host 200.100.14.4 host 224.0.0.10
// 允许 ISAKMP SA 报文通过（SA 使用 UDP 协议，端口号是 500）
R1(config-ext-nacl)#permit udp host 200.100.24.2 host 200.100.14.1 eq 500
// 允许 ESP 加密报文通过
R1(config-ext-nacl)#permit esp host 200.100.24.2 host 200.100.14.1
R1(config-ext-nacl)#exit
R1(config)#interface fa0/0
R1(config-if)#ip access-group outside in
R1(config-if)#exit

8. 测试

测试前，要在路由器 R1、R2、R3 上添加默认路由，路由器是根据路由表转发流量的，若

本路由器中没有数据包要到达的目标网段，数据包将被丢弃。

```
R1(config)#ip route 0.0.0.0 0.0.0.0 200.100.14.4
R2(config)#ip route 0.0.0.0 0.0.0.0 200.100.24.4
R3(config)#ip route 0.0.0.0 0.0.0.0 192.168.13.1
```

1）查看路由器 R1 的 ISAKMP 协议

```
R1#show crypto isakmp policy
Global IKE policy
Protection suite of priority10
        encryption algorithm:           Three key tripleDES
        hashalgorithm:                  Secure Hash Standard
        authentication method:          Pre-Shared Key
        Diffie-Hellman group:           #1 (768 bit)
        lifetime:                       86400seconds，no volume limit
```

2）查看路由器 R1 的 IPsec 变换集

```
R1#show crypto ipsec transform-set
Transform set cxy_lab：{ {esp-3des esp-sha-hmac}
    will negotiate={Tunnel，}，
    Transform set #$!default_transform_set_1：{esp-aes esp-sha-hmac}
    will negotiate={Transport，}，
    Transform set#$!default_transform set_0：{esp-3des esp-sha-hmac
    will negotiate=Transport，}，
}
```

3）查看路由器 R1 的保密映射

```
R1#show crypto map
Crypto Map ipsec_vpn 10 ipsec-isakmp
    Peer=200.100.24.2
    Extended IP access list cxy_vpn
        access-list cxy_vpn permit ip 192.168.3.0 0.0.0.255 192.168.2.0 0.0.0.255
    Current peer：200.100.24.2
    Security association lifetime：4608000 kilobytes / 3600 seconds
    PFS(Y/N)：N
    Transform sets={
        cxy_lab，
    }
    Interfaces using crypto map ipsec_vpn：
```

Fa0/0

4）发送满足条件的数据流，以便 IPsec VPN 加密

（1）创建一个复杂的 PDU 报文。

在 Packet Tracer 的次工具栏中单击"创建复杂的 PDU"（Add Complex PDU）按钮，然后单击路由器 R3 的图标。在弹出的窗口中完成下列操作：在"传出接口"下拉列表框中选择"Loopback0"选项；在"PDU 设置"选区中，在"目的 IP 地址"文本框中输入"192.168.2.1"，在"序列号"文本框中输入"1"；在"模拟设置"选区中，选中"单次"单选按钮，在"时间"文本框中输入"1"；单击"创建 PDU"按钮。

上述操作等效于在路由器 R3 上运行命令"ping 192.168.2.1 source 192.168.3.1"。

（2）创建另一个复杂的 PDU 报文。

同理，创建另一个复杂的 PDU 报文，其操作效果等效于在路由器 R2 上运行命令"ping 192.168.3.1 source 192.168.2.1"。

5）查看 ISAKMP 安全关联

（1）查看路由器 R1 的安全关联。

```
R1#show crypto isakmp sa
IPv4 Crypto ISAKMP SA
dst              src            state       conn-id      slot         status
200.100.24.2     200.100.14.1   QM_DLE      1045         0            ACTIVE
IPv6 Crypto ISAKMP SA
```

（2）查看路由器 R2 的安全关联。

```
R2#show crypto isakmp sa
IPv4 Crypto ISAKMP SA
dst              src            state       conn-id      slot         status
200.100.14.1     200.100.24.2   QM_DLE      1083         0            ACTIVE
IPv6 Crypto ISAKMP SA
```

可见，R1 和 R2 的 ISAKMP 安全关联已经建立。

6）查看满足条件的数据流是否已加密

（1）查看路由器 R1 的加密数据。

```
R1#show crypto ipsec sa
interface：FastEthernet 0/0
    crypto map tag： ipsec_vpn, local addr 200.100.14.1
protected vrf：(none)
local ident (addr/mask/prot/port)：(192.168.3.0/255.255.255.0/0/0)
remote ident (addr/mask/prot/port)：(192.168.2.0/255.255.255.0/0/0)
current_peer 200.100.24.2 port 500
    PERMIT，flags={origin is acl，}
```

#pkts encaps：4，#pkts encrypt：4，#pkts digest：0
#pkts decaps：2，#pkts decrypt：2，#pkts verify：0
#pkts compressed：0，#pkts decompressed：0
#pkts not compressed：0，#pkts compr. failed：0
#pkts not decompressed：0，#pkts decompress failed：0
#send errors 1，#recv errors 0
    local crypto endpt.：200.100.14.1，remote crypto endpt.：200.100.24.2
    path mtu 1500，ip mtu 1500，ip mtu idb FastEthernet 0/0
    current outbound spi：0x77838BF7(2005109751)
    inbound esp sas：
    spi：0xCC10E 507(3423659271)
      transform：esp-3des esp-sha-hmac，
      in use settings={Tunnel，}
      conn id：2001，flow_id：FPGA：1，crypto map：ipsec_vpn
      sa timing：remaining key lifetime (k/sec)：(4525504/1807)
      IV size：16 bytes
      replay detection support：N
      Status：ACTIVE
    inbound ah sas：
    inbound pcp sas：
    outbound esp sas：
     spi：0x77838BF7(2005109751)
      transform：esp-3desesp-sha-hmac，
      in use settings={Tunnel，}
      conn id：2002, flow_id：FPGA：1，crypto map：ipsec_vpn
      sa timing：remaining key lifetime(k/sec)：(4525504/1807)
      IV size：16 bytes
      replay detection support：N
      status：ACTIVE

（2）查看路由器 R2 的加密数据。

R2#show crypto ipsec sa
interface：FastEthernet 0/0
    crypto map tag：ipsec_vpn，local addr 200.100.24.2
protected vrf：(none)
local ident (addr/mask/prot/port)：(192.168.2.0/255.255.255.0/0/0)
remote ident (addr/mask/prot/port)：(192.168.3.0/255.255.255.0/0/0)
current_peer 200.100.14.1 port 500

　　　　PERMIT，flags={origin is acl，}
　#pkts encaps：69，#pkts encrypt：69，#pkts digest：0
　#pkts decaps：69，#pkts decrypt：69，#pkts verify：0
　#pkts compressed：0，#pkts decompressed：0
　#pkts not compressed：0，#pkts compr. failed：0
　#pkts not decompressed：0，#pkts decompress failed：0
　#send errors 0，#recv errors 0
　　　local crypto endpt.：200.100.24.2，remote crypto endpt.：200.100.14.1
　　　path mtu 1500，ip mtu 1500，ip mtu idb FastEthernet 0/0
　　　current outbound spi：0xCC10E 507(3423659271)
　　　inbound esp sas：
　　　spi：0x77838BF7(2005109751)
　　　　transform：esp-3des esp-sha-hmac，
　　　　in use settings={Tunnel，}
　　　　conn id：2001，flow_id：FPGA：1，crypto map：ipsec_vpn
　　　　sa timing：remaining key lifetime (k/sec)：(4525504/730)
　　　　IV size：16 bytes
　　　　replay detection support：N
　　　　Status：ACTIVE
　　　inbound ah sas：
　　　inbound pcp sas：
　　　outbound esp sas：
　　　spi：0xCC10E 507(3423659271)
　　　　transform：esp-3desesp-sha-hmac，
　　　　in use settings={Tunnel，}
　　　　conn id：2002, flow_id：FPGA：1，crypto map：ipsec_vpn
　　　　sa timing：remaining key lifetime(k/sec)：(4525504/730)
　　　　IV size：16 bytes
　　　　replay detection support：N
　　　　status：ACTIVE

9. 抓包分析

1）创建复杂的 PDU

创建复杂的 PDU，等效于在路由器 R3 上运行命令"ping 192.168.2.1 source 192.168.3.1"。

2）单击状态栏右下角处的"Simulation"按钮，弹出"模拟面板"窗口（见图 11-2）

第 11 章 基于路由器的 IPsec VPN

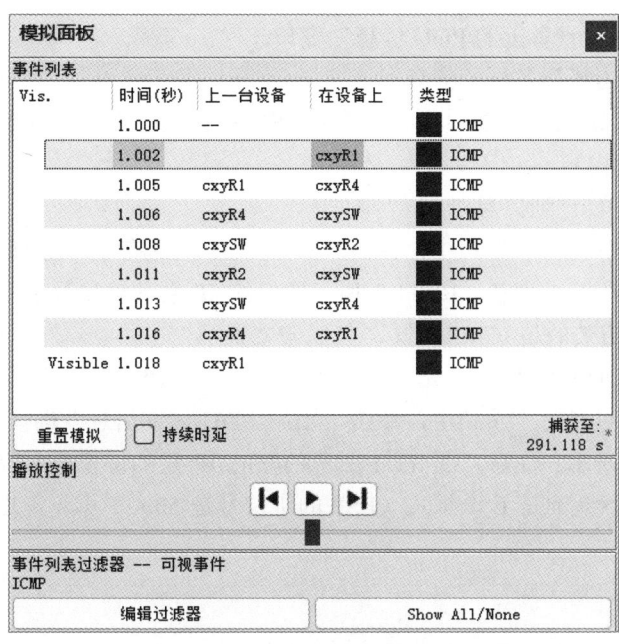

图 11-2 "模拟面板"窗口

3）在图 11-3 的"播放控制"选区中，单击播放图标，开始进行模拟

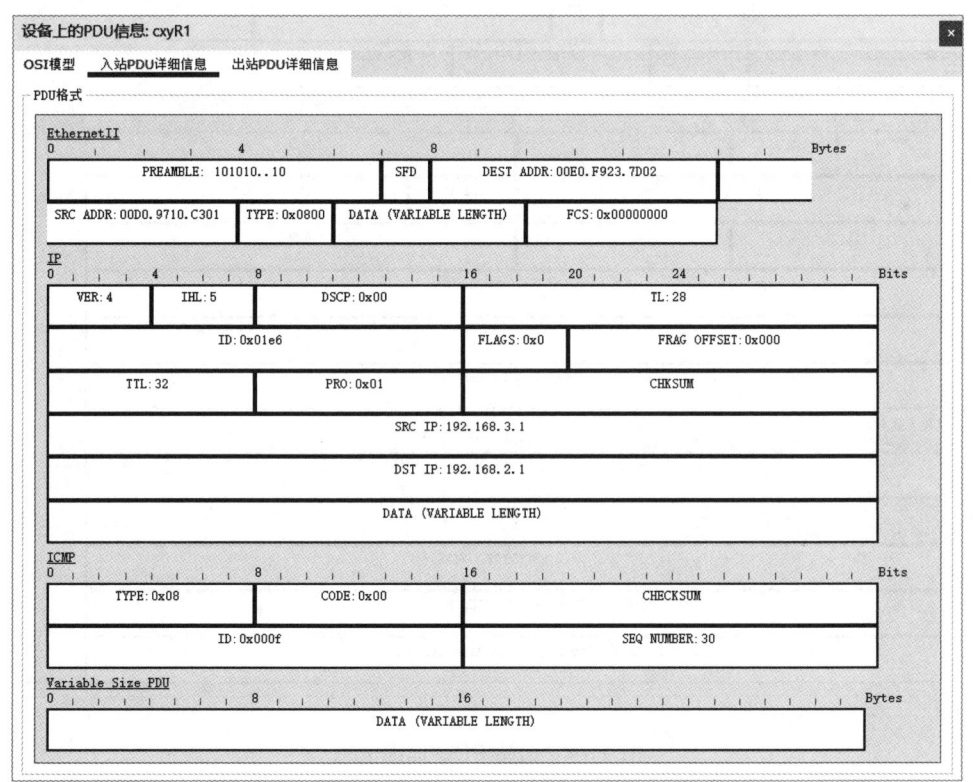

图 11-3 入站 PDU 详细信息

4）依次抓包并分析

模拟结束后，在"模拟面板"窗口的"事件列表"中将显示若干个报文。单击"事件列表"

中的第二个报文，弹出"设备上的 PDU 信息"窗口。

（1）显示 OSI 参考模型各层的信息。

在"设备上的 PDU 信息"窗口中，单击"OSI 模型"选项卡，将从低层到高层逐层显示 OSI 参考模型中各层的内容。

（2）显示入站 PDU 详细信息。

在"设备上的 PDU 信息"窗口中，单击"入站 PDU 详细信息"选项卡，将显示入站 PDU 详细信息，如图 11-3 所示。注意，图 11-3 所示的帧的 IP 层的协议号字段的内容是"0x01"，意味着该帧的 IP 分组封装的是 ICMP 报文，且以明文表示。

（3）显示出站 PDU 详细信息。

在"设备上的 PDU 信息"窗口中，单击"出站 PDU 详细信息"选项卡，将显示出站 PDU 详细信息，如图 11-4 所示。注意，图 11-4 所示的帧的 IP 层的协议号字段的内容是"0x32"，意味着该帧的 IP 分组封装的是 ESP 报文（ESP 的协议号是 50）。可见，原来的 ICMP 报文现在以密文表示。

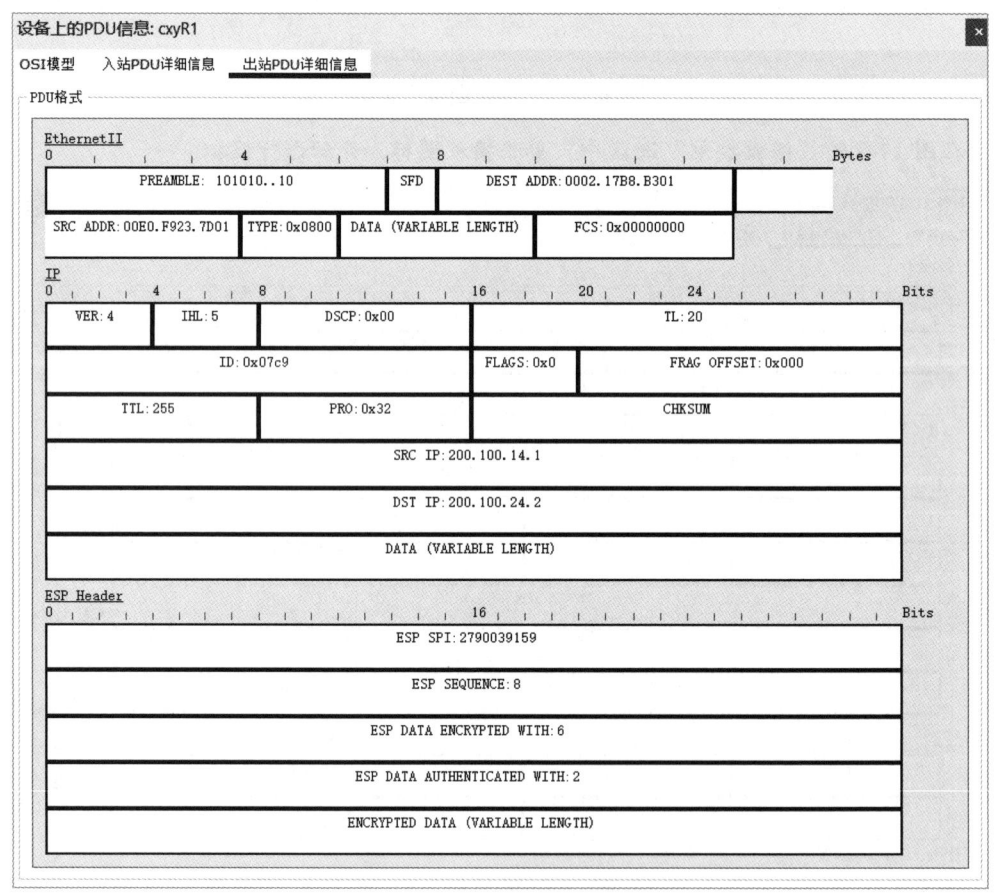

图 11-4 出站 PDU 详细信息

类似地，可以对其他报文进行分析。

第 12 章
基于路由器的 Easy VPN

12.1 组网需求

网络拓扑如图 12-1 所示,公司总部的内网连接在路由器 R1 的 Fa0/1 接口上,R1 通过 Fa0/0 连接到互联网,互联网是由 R2、R4 和交换机 S0 组成的网络模拟的。现要求将路由器 R1 配置成 Easy VPN 网关（VPN 服务器）,其中加密算法选择 3DES,Hash 函数使用 SHA,设备认证使用预共享密钥。在 VPN 网关上启用 AAA（Authentication,Authorization,and Accounting,认证授权、计费）功能,并配置认证方式为本地,定义在本地进行授权。当公司员工出差到外地后,可以在笔记本电脑上打开 Easy VPN 客户端,通过互联网拨号接入 Easy VPN 服务器,进而访问公司内部网络资源。例如,浏览内部 Web 服务器上的网页,下载内部文件服务器上的文件,等等。

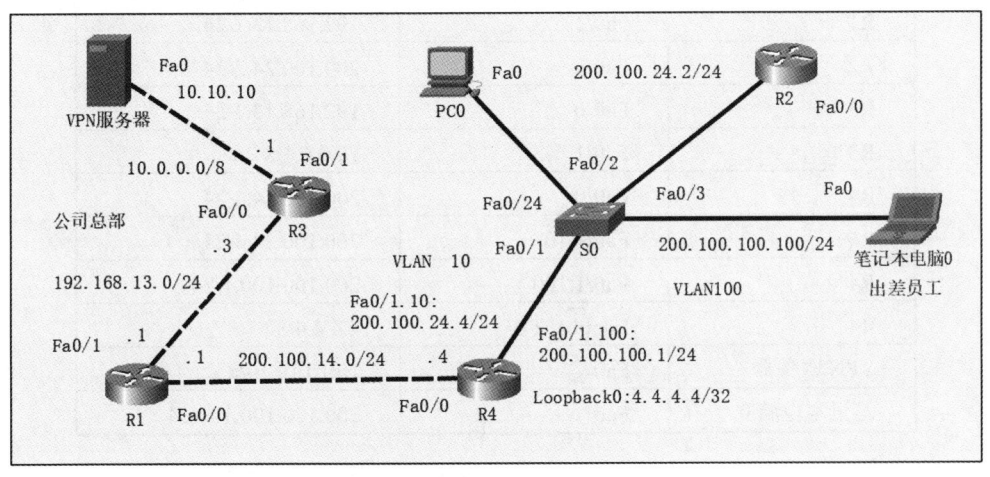

图 12-1 网络拓扑

12.2 仿真实验

12.2.1 仿真环境

仿真环境所需要的硬件、软件建议至少满足以下要求。

- 物理机的处理器主频为 1.6GHz;

- 内存 RAM 大小为 32.0GB；
- 物理机的操作系统为 Windows 10；
- 虚拟机为 VMware Workstation 16 Pro；
- 网络模拟器选用 Cisco Packet Tracer，版本 8.0。

12.2.2 实验设计

根据网络拓扑（图 12-1）完成网络设备的选型：4 台路由器选用 Cisco 2811 系列路由器，1 台交换机选用 Cisco WS-C2950-24 系列二层交换机，服务器选用通用型服务器 Server-PT。根据前面的组网需求，本实验设计如下。

1. VLAN 设计

在交换机上新建两个 VLAN：VLAN10 和 VLAN100，将接口 Fa0/2 和 Fa0/24 划分到 VLAN10 中，将接口 Fa0/3 划分到 VLAN100 中。将接口 Fa0/1 设置为 trunk 模式并允许所有 VLAN 通过。

2. IP 地址规划

网络中有 5 个物理子网，网络地址分别为 200.100.14.0/24、192.168.13.0/24、200.100.24.0/24、10.0.0.0/8 和 200.100.100.0/24。此外，还有 1 个环回地址。各设备接口的 IP 地址/掩码如表 12-1 所示。

表 12-1 各设备接口的 IP 地址/掩码

设备	接口	IP 地址/掩码
R1	Fa0/0	200.100.14.1/24
R1	Fa0/1	192.168.13.1/24
R2	Fa0/0	200.100.24.2/24
R3	Fa0/0	192.168.13.3/24
R3	Fa0/1	10.0.0.1/8
R4	Fa0/0	200.100.14.4/24
R4	Fa0/1.10	200.100.24.4/24
R4	Fa0/1.100	200.100.100.1/24
R4	Loopback 0	4.4.4.4/32
VPN 服务器	Fa0	10.10.10.10/8
笔记本电脑 0	Fa0	200.100.100.100/24

3. 路由设计

在公网上配置 EIGRP 协议，使得公网中的所有 IP 地址在网络层互通：在 R1、R2 和 R4 上分别启用 EIGRP 协议，AS 号均为 100，取消自动汇总。R1 在 200.100.14.0/24 上声明网段，R2 在 200.100.24.0/24 上声明网段，R4 在 200.100.14.0/24、200.100.24.0/24 和 200.100.100.0/24 上声明网段。

在内网上配置 RIP 协议，使得内网中的所有 IP 地址在网络层互通：在 R1 和 R3 上分别启用 RIPv2 协议，取消自动汇总。R1 在 192.168.13.0 上声明网段，R3 在 192.168.13.0 和 10.0.0.0 上声明网段。

4. Easy VPN 设计

在路由器 R1 上配置 Easy VPN 服务器，具体包括在路由器 R1 上开启并配置 ISAKMP 协议，加密算法选择 3DES，Hash 函数使用 SHA，认证使用预共享密钥。在 R1 上定义地址池（如 192.168.100.1～192.168.100.100），用于给远程接入的 VPN 客户端分配 IP 地址。在路由器 R1 上启用 AAA 功能，并配置认证方式为本地，定义在本地进行授权。

当出差员工希望通过 Easy VPN 客户端访问内网资源时，首先在笔记本电脑上运行 VPN 客户端，然后按要求输入用户名及密码。VPN 连接成功后，在笔记本电脑中会自动增加一个隧道接口，此 VPN 隧道的 IP 地址将从 VPN 服务器的地址池中分配（如 192.168.100.2）。同时，在 VPN 服务器上也会相应地自动增加一条静态路由。连接成功后，出差员工就可以访问公司总部的内部资源（如内部 Web 服务器、内部 FTP 服务器）了。

12.2.3 操作步骤

下面给出本次实验的具体操作，一共包括 7 个步骤。

1. 按照图 12-1 所示的网络拓扑，在 PT 中新建一个空白的工程

4 台路由器选用 Cisco 2811 系列路由器，1 台交换机选用 Cisco WS-C2950-24 系列二层交换机，服务器选用 PT 通用型服务器 Server-PT。

2. 配置交换机

按照以下内容，对交换机 S0 进行相应的配置。注意：加粗的命令是需要进行配置的。

```
Switch(config)#hostname cxyS0
cxyS0#show running-config
Building configuration...

Current configuration : 1256 bytes
!
version 15.0
no service timestamps log datetime msec
no service timestamps debug datetime msec
no service password-encryption
!
hostname cxyS0            // 无缩进表示在全局配置模式下运行此命令
!
spanning-tree mode pvst
spanning-tree extend system-id
!
interface FastEthernet0/1
 switchport mode trunk    // 缩进表示在接口子配置模式下运行此命令
!
interface FastEthernet0/2
```

```
   switchport access vlan10
   switchport mode access
!
interface FastEthernet0/3
  switchport access vlan100
  switchport mode access
!
interface FastEthernet0/4
!
interface FastEthernet0/5
!
// 此处省略了若干行
!
interface FastEthernet0/23
!
interface FastEthernet0/24
  switchport access vlan10
  switchport mode access
!
interface GigabitEthernet0/1
!
interface GigabitEthernet0/2
!
interface Vlan1
 no ip address
 shutdown
!
line con 0
!
line vty 0 4
 login
line vty 5 15
 login
!
End
```

3. 配置路由器接口的 IP 地址

1）配置路由器 **R1** 接口的 **IP** 地址

R1(config)#interface fa0/0
R1(config-if)#ip address 200.100.14.1 255.255.255.0

R1(config-if)#no shutdown
R1(config-if)#exit
R1(config)#interface fa0/1
R1(config-if)#ip address 192.168.13.1 255.255.255.0
R1(config-if)#no shutdown

2）配置路由器 R2 接口的 IP 地址

R2(config)#interface fa0/0
R2(config-if)#ip address 200.100.24.2 255.255.255.0
R2(config-if)#no shutdown

3）配置路由器 R3 接口的 IP 地址

R3(config)#interface fa0/0
R3(config-if)#ip address 192.168.13.3 255.255.255.0
R3(config-if)#no shutdown

4）配置路由器 R4 接口的 IP 地址

R4(config)#interface fa0/0
R4(config-if)#ip address 200.100.14.4 255.255.255.0
R4(config-if)#no shutdown
R4(config-if)#exit
R4(config)#interface fa0/1
R4(config-if)#no shutdown
R4(config-if)#exit
R4(config)#interface fa0/1.10
R4(config-subif)#encapsulation dot1Q 10
R4(config-subif)#ip address 200.100.24.4 255.255.255.0
R4(config-subif)#exit
R4(config)#interface fa0/1.100
R4(config-subif)#encapsulation dot1Q 100
R4(config-subif)#ip address 200.100.100.1 255.255.255.0
R4(config-subif)#exit
R4(config)#interface loopback0
R4(config-if)#ip address 4.4.4.4 255.255.255.255
R4(config-if)#exit

5）测试直连网段的连通性

R3#ping 192.168.13.1
Type escape sequence to abort.
Sending 5, 100-byte ICMP Echos to 192.168.13.1, timeout is 2 seconds:

.!!!!

Success rate is 80 percent (4/5), round-trip min/avg/max = 1/2/4 ms

R1#ping 200.100.14.4

Type escape sequence to abort.

Sending 5, 100-byte ICMP Echos to 200.100.14.4, timeout is 2 seconds:

.!!!!

Success rate is 80 percent (4/5), round-trip min/avg/max = 1/1/4 ms

R4#ping 200.100.24.2

Type escape sequence to abort.

Sending 5, 100-byte ICMP Echos to 200.100.24.2, timeout is 2 seconds:

.!!!!

Success rate is 80 percent (4/5), round-trip min/avg/max = 1/2/4 ms

4. 在公网路由器上配置 EIGRP 协议

1）在 R1 上配置 EIGRP 协议

```
R1(config)#router eigrp 100
R1(config-router)#no auto-summary
R1(config-router)#network 200.100.14.1 0.0.0.0
R1(config-router)#exit
```

2）在 R2 上配置 EIGRP 协议

```
R2(config)#router eigrp 100
R2(config-router)#no auto-summary
R2(config-router)#network 200.100.24.2 0.0.0.0
R2(config-router)#exit
```

3）在 R4 上配置 EIGRP 协议

```
R4(config)#router eigrp 100
R4(config-router)#no auto-summary
R4(config-router)#network 200.100.14.4 0.0.0.0
R4(config-router)#network 200.100.24.4 0.0.0.0
R4(config-router)#network 200.100.100.1 0.0.0.0
R4(config-router)#exit
```

5. 在私网路由器上配置 RIP 协议

1）在 R1 上配置 RIP 协议

```
R1(config)#router rip
R1(config-router)#version 2
R1(config-router)#no auto-summary
R1(config-router)#network 192.168.13.0
```

// 把静态路由发布给路由器 R3，静态路由包含了 VPN 客户端连通后自动产生的主机路由
R1(config-router)#redistribute static
R1(config-router)#exit

2）在 R3 上配置 RIP 协议

R3(config)#router rip
R3(config-router)#version 2
R3(config-router)#no auto-summary
R3(config-router)#network 192.168.13.0
R3(config-router)#network 10.0.0.0
R3(config-router)#exit

6. 在路由器 R1 上配置 Easy VPN
1）配置 AAA

R1(config)#aaa new-model   // 启用 AAA 功能
R1(config)#aaa authentication login cxyVPNuserAuthen local   // 定义一个名为 "cxyVPNuserAuthen"
                                                            // 的认证，认证方式为本地
R1(config)#aaa authorization network ezvpn-author local   // 定义在本地进行授权
R1(config)#username cxyuser password cxyuser   // 定义一个用户名和密码

2）开启和配置 ISAKMP 协议

R1(config)#crypto isakmp policy 10
R1(config-isakmp)#encryption 3des
R1(config-isakmp)#hash sha
R1(config-isakmp)#group 2
R1(config-isakmp)#authentication pre-share
R1(config-isakmp)#exit

3）定义地址池

R1(config)#ip local pool ezvpn-pool 192.168.100.1 192.168.100.100

4）创建组及组策略

R1(config)#crypto isakmp client configuration group ezvpn   // 创建一个组策略
R1(config-isakmp-group)#key CXY123   // 设置组密码
R1(config-isakmp-group)#pool ezvpn-pool
R1(config-isakmp-group)#exit

5）定义变换集

R1(config)#crypto ipsec transform-set cxy_lab esp-3des esp-sha-hmac
R1(cfg-crypto-trans)#exit

6）定义保密映射

R1(config)#crypto dynamic-map dynmap 10
R1(config-crypto-map)#set transform-set cxy_lab
R1(config-crypto-map)#reverse-route
R1(config-crypto-map)#exit
R1(config)#crypto map cxyClientMap 10 ipsec-isakmp dynamic DYNMAP
R1(config)#crypto map cxyClientMap isakmp authorization list ezvpn-author
R1(config)#crypto map cxyClientMap client authentication list cxyVPNuserAuthen
R1(config)#crypto map cxyClientMap client configuration address respond
R1(config)#crypto isakmp client configuration group ezvpn
cxyR1(config-isakmp-group)#key CXY123
cxyR1(config-isakmp-group)#pool ezvpn-pool

7）接口应用保密映射

R1(config)#interface fa0/0
R1(config-if)#crypto map cxyClientMap
R1(config-if)#exit

7. 笔记本电脑通过 Easy VPN 客户端访问内网资源

1）在笔记本电脑上运行 VPN 客户端

（1）启动 VPN 客户端。

在 PT 工作区窗口，单击"笔记本电脑 0"图标，弹出如图 12-2 所示的窗口。在"桌面"选项中，单击"VPN"图标以启动 VPN 客户端，弹出如图 12-3 所示的"VPN 配置"窗口。

图 12-2　启动 VPN 客户端

（2）输入 VPN 用户名及密码。

在图 12-3 所示的"VPN 配置"窗口的"组名"文本框中输入"ezvpn"，在"组密钥"文本框中输入"CXY123"，在"主机 IP（服务器 IP）"文本框中输入"200.100.14.1"，在"用户名"文本框中输入"cxyuser"，在"密码"文本框中输入"cxyuser"，单击"连接"按钮，创建新的 VPN 连接。

图 12-3 "VPN 配置"窗口

（3）VPN 连接成功。

VPN 连接成功后，将显示如图 12-4 所示的窗口。若需要退出，则单击"中断连接"按钮。

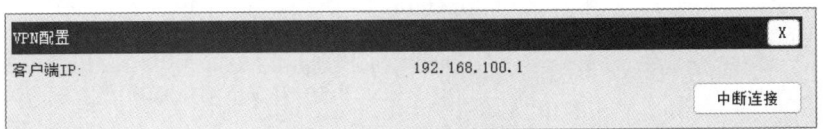

图 12-4 VPN 连接成功

（4）笔记本电脑中增加了一个隧道接口。

VPN 连接成功后，在笔记本电脑中会自动增加一个隧道接口，此 VPN 隧道的 IP 地址是 192.168.100.2。

在 PT 的"命令提示符"窗口中可以查看隧道接口的 IP 地址：

```
Packet Tracer PC Command Line 1.0
C:\>ipconfig /all

FastEthernet0 Connection:(default port)

Connection-specific DNS Suffix  :
Physical Address                : 0090.2144.7945
Link-local IPv6 Address         : FE80::290:21FF:FE44:7945
IPv6 Address                    : ::
IPv4 Address                    : 200.100.100.100
Subnet Mask                     : 255.255.255.0
Default Gateway                 : ::
```

	200.100.100.1
DHCP Servers	: 0.0.0.0
DHCPv6 IAID	
DHCPv6 Client DUID	: 00-01-00-01-59-76-5D-90-00-90-21-44-79-45
DNS Servers	: ::
	0.0.0.0

Bluetooth Connection:

Connection-specific DNS Suffix	:
Physical Address	: 000A.415C.165E
Link-local IPv6 Address	: ::
IPv6 Address	: ::
IPv4 Address	: 0.0.0.0
Subnet Mask	: 0.0.0.0
Default Gateway	: ::
	0.0.0.0
DHCP Servers	: 0.0.0.0
DHCPv6 IAID	:
DHCPv6 Client DUID	: 00-01-00-01-59-76-5D-90-00-90-21-44-79-45
DNS Servers	: ::
	0.0.0.0

Tunnel Interface IP Address: 192.168.100.2

（5）通过 VPN 认证后，路由器 R1 中增加了一条静态路由。

```
R1#show ip route
Codes: L-local, connected, S-static, R-RIP, M-mobile, B-BGP
       D_EIGRP, EX-EIGRP external, O-OSPF, IA-OSPF inter area
       N1-OSPF NSSA external type 1, N2-OSPF NSSA external type 2
       E1-OSPF external type 1, E2-OSPF external type 2, E-EGP
       i-IS-IS, L1-IS-IS level-1, L2-IS-IS level-2, ia — IS-IS inter area
       *-candidate default, U-per-user static route, o-ODR
       P-periodic downloaded static route
  Gateway of last resort is not set
  R  10.0.0.0/8 [120/1] via 192.168.13.3, 00: 00: 10, FastEthernet 0/1
192.168.13.0/24 is variably subnetted, 2 subnets, 2 masks
  C  192.168.13.0/24 is directly connected, FastEthernet 0/1
  L  192.168.13.1/32 is directly connected, FastEthernet 0/1
```

192.168.100.0/32 is subnetted, 3 subnets
S   192.168.100.4/32 [1/0] via 200.100.100.100
S   192.168.100.5/32 [1/0] via 200.100.100.100
S   192.168.100.6/32 [1/0] via 200.100.100.100
200.100.14.0/24 is variably subnetted, 2 subnets, 2 masks
C   200.100.14.0/24 is directly connected, FastEthernet 0/0
L   200.100.14.1/32 is directly connected, FastEthernet 0/0
D   200.100.24.0/24 [90/30720] via 200.100.14.4, 06:42:00, FastEthernet 0/0
D   200.100.100.0/24 [90/30720] via 200.100.14.4, 06:42:00, FastEthernet 0/0

（6）通过 VPN 认证后，路由器 R3 通过 RIP 学习获得了一条新增路由。

R3#show ip route
Codes：L-local, connected, S-static, R-RIP, M-mobile, B-BGP
       D_EIGRP, EX-EIGRP external, O-OSPF, IA-OSPF inter area
       N1-OSPF NSSA external type 1, N2-OSPF NSSA external type 2
       E1-OSPF external type 1, E2-OSPF external type 2, E-EGP
       i-IS-IS, L1-IS-IS level-l, L2-IS-IS level-2, ia — IS-IS inter area
       *-candidate default, U-per-user static route, o-ODR
       P-periodic downloaded static route
Gateway of last resort is not set
    10.0.0.0/8 is variably subnetted, 2 subnets, 2 masks
C   10.0.0.0/8 is directly connected, FastEthernet 0/1
L   10.0.0.1/32 is directly connected, FastEthernet 0/1
    192.168.13.0/24 is variably subnetted, 2 subnets, 2 masks
C   192.168.13.0/24 is directly connected, FastEthernet 0/0
L   192.168.13.3/32 is directly connected, FastEthernet 0/0
    192.168.100.0/32 is subnetted, 3 subnets
R   192.168.100.4/32 [120/1] via 192.168.13.1, 00:00:26, FastEthernet 0/0
R   192.168.100.5/32 [120/1] via 192.168.13.1, 00:00:26, FastEthernet 0/0
R   192.168.100.6/32 [120/1] via 192.168.13.1, 00:00:26, FastEthernet 0/0

2）出差员工访问公司总部的内部资源

连接成功后，出差员工就可以访问公司总部的内部资源了，下面以访问 Web 服务器和 FTP 服务器为例进行说明。

（1）访问 Web 服务器（图 12-5）。

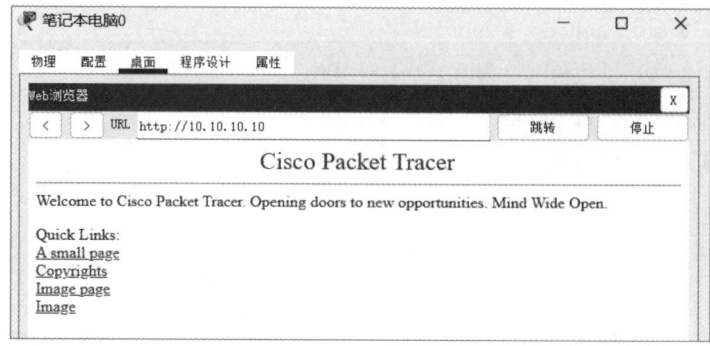

图 12-5　出差员工通过 VPN 访问 Web 服务器

（2）访问 FTP 服务器。

VPN 连接成功后，出差员工还可以访问 FTP 服务器下载文件。

C:\>ftp 10.10.10.10
Trying to connect...10.10.10.10
Connected to 10.10.10.10
220- Welcome to PT Ftp server
Username:cisco
331- Username ok, need password
Password:
230- Logged in
(passive mode On)
ftp>dir

Listing /ftp directory from 10.10.10.10:
0 : asa842-k8.bin 5571584
1 : asa923-k8.bin 30468096
2 : c1841-advipservicesk9-mz.124-15.T1.bin 33591768
3 : c1841-ipbase-mz.123-14.T7.bin 13832032
4 : c1841-ipbasek9-mz.124-12.bin 16599160
5 : c1900-universalk9-mz.SPA.155-3.M4a.bin 33591768
……
ftp>get asa842-k8.bin

Reading file asa842-k8.bin from 10.10.10.10:
File transfer in progress...

[Transfer complete - 5571584 bytes]

5571584 bytes copied in 9.241 secs (138147 bytes/sec)
ftp>

第 4 篇　防火墙安全篇

第 13 章　ASA 防火墙的基本配置

第 14 章　ASA 防火墙的路由配置

第 15 章　ASA 防火墙的 NAT 和 ACL 配置

第 16 章　策略路由

第 17 章　基于 ASA 防火墙的 IPsec VPN 配置

第 13 章 ASA 防火墙的基本配置

13.1 组网需求

网络拓扑如图 13-1 所示,其中 ASA 为 Cisco ASA 系列防火墙,现要求在防火墙上实现以下功能。

(1) 定义防火墙的安全区域及安全等级;
(2) 配置防火墙的域名和时间;
(3) 开启日志功能,设置为本地保存,级别为 informational,并开启时间标识;
(4) 设置允许用户通过 Telnet 远程登录防火墙;
(5) 设置要求用户只能通过 SSH 安全远程登录防火墙进行管理;
(6) 配置模块化策略框架(Modular Policy Framework,MPF),实现内网中的路由器 R2 可以 ping 通 DMZ 区域中的路由器 R3。

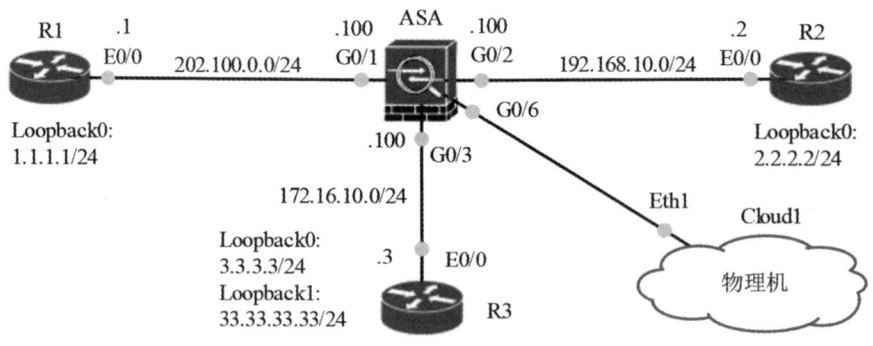

图 13-1 网络拓扑

13.2 仿真实验

13.2.1 仿真环境

仿真环境所需要的硬件、软件建议至少满足以下要求。

- 物理机的处理器主频为 1.6GHz;
- 内存 RAM 大小为 32.0GB;
- 物理机的操作系统为 Windows 10;

- 虚拟机为 VMware Workstation 16 Pro；
- 网络模拟器选用 GNS3 2.2.31；
- 防火墙为 Cisco ASAv9.17.1；
- 路由器为 Cisco IOL Router15.7。

13.2.2 实验设计

根据网络拓扑（图 13-1）完成网络设备的选型：防火墙选用 Cisco ASAv9.17.1，3 台路由器选用 Cisco IOL Router15.7 系列路由器，物理机通过 Cloud1 的 Eth1 接口接入 GNS3 网络中。根据前面的组网需求，本实验设计如下。

1. IP 地址规划

网络有 3 个物理子网，网络地址分别为 202.100.0.0/24、192.168.10.0/24、172.16.10.0/24，此外还有 4 个环回地址。各设备接口的 IP 地址/掩码如表 13-1 所示。

表 13-1 各设备接口的 IP 地址/掩码

设备	接口	IP 地址/掩码
R1	E0/0	202.100.0.1/24
R1	Loopback0	1.1.1.1/24
R2	E0/0	192.168.10.2/24
R2	Loopback0	2.2.2.2/24
R3	E0/0	172.16.10.3/24
R3	Loopback0	3.3.3.3/24
R3	Loopback1	33.33.33.33/24
ASA	G0/1	202.100.0.100/24
ASA	G0/2	192.168.10.100/24
ASA	G0/3	172.16.10.100/24

2. 路由设计

在 3 台路由器上分别配置默认路由，R1 的下一跳 IP 地址是 202.100.0.100，R2 的下一跳 IP 地址是 192.168.10.100，R3 的下一跳 IP 地址是 172.16.10.100。

3. 防火墙设计

将 ASA 的接口 G0/1 定义为外部接口，安全级别采用默认值 0；将接口 G0/2 定义为内部接口，安全级别采用默认值 100；将接口 G0/3 定义为 DMZ 接口，安全级别设置为 50。

开启防火墙的日志功能，级别为 informational，设置为本地保存。配置 SSH，只允许管理员从 192.168.10.2 通过 SSH 安全远程登录 ASA。

配置 MPF，MPF 默认禁止 ICMP 数据包通过，故当内网主机 ping 外网主机时，外网主机的应答报文会被 ASA 的外部接口拒绝。也就是说，外网的 ICMP 报文是进入不了内网的。因此，需要在默认的 MPF global_policy 中设置允许 ICMP 报文进入，使内网 R2 可以 ping 通 R1。

此外，为了提高配置文件的安全性和可靠性，将防火墙 ASA 的配置文件导出到远程的 FTP 服务器上。

13.2.3 操作步骤

下面给出本次实验的具体操作，一共包括 11 个步骤。

1. 按照图 13-1 所示的网络拓扑，在 GNS3 中新建一个空白的工程

防火墙选用 Cisco ASAv9.17.1 系列防火墙，3 台路由器选用 Cisco IOL Router15.7 系列路由器。

物理机通过 Cloud1 的 Eth1 接口（对应虚拟网卡的 VMware Network Adapter VMnet8 网络，IP 地址为 192.168.241.1/24）接入 GNS3 网络中。

2. 配置路由器的 IP 地址

1）配置路由器 R1 的 IP 地址

```
R1(config)#hostname cxyR1
cxyR1(config)#interface e0/0
cxyR1(config-if)#ip address 202.100.0.1 255.255.255.0
cxyR1(config-if)#no shutdown
cxyR1(config-if)#exit
cxyR1(config)#interface loopback0
cxyR1(config-if)#ip address 1.1.1.1 255.255.255.0
cxyR1(config-if)#exit
```

2）配置路由器 R2 的 IP 地址

```
R2(config)#hostname cxyR2
cxyR2(config)#interface e0/0
cxyR2(config-if)#ip address 192.168.10.2 255.255.255.0
cxyR2(config-if)#no shutdown
cxyR2(config-if)#exit
cxyR2(config)#interface loopback0
cxyR2(config-if)#ip address 2.2.2.2 255.255.255.0
cxyR2(config-if)#exit
```

3）配置路由器 R3 的 IP 地址

```
R3(config)#hostname cxyR3
cxyR3(config)#interface e0/0
cxyR3(config-if)#ip address 172.16.10.3 255.255.255.0
cxyR3(config-if)#no shutdown
cxyR3(config-if)#exit
cxyR3(config)#interface loopback0
cxyR3(config-if)#ip address 3.3.3.3 255.255.255.0
cxyR3(config-if)#exit
cxyR3(config)#interface loopback1
```

cxyR3(config-if)#ip address 33.33.33.33 255.255.255.0
cxyR3(config-if)#exit

3. 配置 3 台路由器的默认路由

cxyR1(config)#ip route 0.0.0.0 0.0.0.0 202.100.0.100
cxyR2(config)#ip route 0.0.0.0 0.0.0.0 192.168.10.100
cxyR3(config)#ip route 0.0.0.0 0.0.0.0 172.16.10.100

4. 配置 ASA 接口地址

ciscoasa> enable	//进入特权模式
Enter Password: cxy123cxy	//设置密码，至少 8 位
Repeat Password: cxy123cxy	//记住该密码，以后登录会用到
ciscoasa#configure terminal	//进入全局配置模式
ciscoasa(config)#hostname cxyASA	//配置主机名
cxyASA(config)#	
cxyASA(config)#interface g0/1	
cxyASA(config-if)#ip address 202.100.0.100 255.255.255.0	
cxyASA(config-if)#nameif outside	//将该接口定义为外部接口
INFO: Security level for "outside" set to 0 by default	//外部接口默认安全级别为 0
cxyASA(config-if)#no shutdown	
cxyASA(config-if)#exit	
cxyASA(config)#interface g0/2	
cxyASA(config-if)#ip address 192.168.10.100 255.255.255.0	
cxyASA(config-if)#nameif inside	//将该接口命名为 inside 接口
INFO: Security level for "inside" set to 100 by default.	
cxyASA(config-if)#no shutdown	
cxyASA(config-if)#exit	
cxyASA(config)#interface g0/3	
cxyASA(config-if)#ip address 172.16.10.100 255.255.255.0	
cxyASA(config-if)#nameif DMZ	//将该接口定义为 DMZ 接口
INFO: Security level for "DMZ" set to 0 by default.	//DMZ 默认安全级别为 0
cxyASA(config-if)#security-level 50	//更改 DMZ 接口的安全级别为 50
cxyASA(config-if)#no shutdown	
cxyASA(config-if)#exit	

5. 查看 ASA 的接口信息
1）查看 ASA 的接口状态

cxyASA(config)#show interface ip brief

Interface	IP-Address	OK?	Method	Status	Protocol
GigabitEthernet0/0	unassigned	YES	unset	administratively down	down
GigabitEthernet0/1	202.100.0.100	YES	manual	up	up
GigabitEthernet0/2	192.168.10.100	YES	manual	up	up
GigabitEthernet0/3	172.16.10.100	YES	manual	up	up
GigabitEthernet0/4	unassigned	YES	unset	administratively down	down
GigabitEthernet0/5	unassigned	YES	unset	administratively down	down
GigabitEthernet0/6	unassigned	YES	unset	administratively down	down
Internal-Data 0/0	169.254.1.1	YES	unset	up	up
Management 0/0	unassigned	YES	unset	administratively down	down

2）查看 ASA 的 IP 地址及接口名

cxyASA(config)#show ip address   //show 命令可以在全局配置模式中运行
System IP Addresses：

Interface	Name	IP_address	Sub netmask	Method
GigabitEthernet 0/1	outside	202.100.0.100	255.255.255.0	manual
GigabitEthernet 0/2	inside	192.168.10.100	255.255.255.0	manual
GigabitEthernet 0/3	DMZ	172.16.10.100	255.255.255.0	manual

Current IP Addresses：

InterfaceName	Name	IP_address	Sub netmask	Method
GigabitEthernet 0/1	outside	202.100.0.100	255.255.255.0	manual
GigabitEthernet 0/2	inside	192.168.10.100	255.255.255.0	manual
GigabitEthernet 0/3	DMZ	172.16.10.100	255.255.255.0	manual

3）测试连通性

cxyASA(config)#ping 202.100.0.1
Type escape sequence to abort.
Sending 5，100-byte ICMP Echos to 202.100.0.1，timeout is 2 seconds：
!!!!!
Success rate is 100 percent(5/5)，round-trip min/avg/max=10/10/10ms

cxyASA(config)#ping 192.168.10.2
Type escape sequence to abort.
Sending 5，100-byte ICMP Echos to 192.168.10.2，timeout is 2 seconds：
!!!!!
Success rate is 100 percent(5/5)，round-trip min/avg/max=1/10/20ms

cxyASA(config)#ping 172.16.10.3
Type escape sequence to abort.

Sending 5，100-byte ICMP Echos to 172.16.10.3，timeout is 2 seconds：

!!!!!

Success rate is 100 percent(5/5)，round-trip min/avg/max=1/8/10ms

6. 配置 ASA 的域名和时间

cxyASA(config)#domain-name cxy.cn
cxyASA(config)#clock set 08:08:08 20 october 2022   //修改成当前日期及时间
cxyASA(config)#show clock   //查看时间
08:08:16.709 UTC Thu Oct 20 2023

7. 设置日志

1）开启日志功能，设置为本地保存，级别为 informational

cxyASA(config)#logging enable
cxyASA(config)#logging buffered informational
cxyASA(config)#logging timestamp

2）查看日志

cxyASA(config)#show logging
Syslog logging：enabled
Facility：20
Timestamp logging：enabled
Timezone：disabled
Hide Username logging：enabled
Standby logging：disabled
Debug-tracelogging：disabled
Console logging：disabled
Monitor logging：disabled
Buffer logging：level informational，5 messages logged
Trap logging：disabled
Permit-hostdown logging：disabled
History logging：disabled
Device ID：disabled
Mail logging：disabled
ASDM logging：disabled
%ASA-5-111008：User 'enable_15' executed the 'logging buffered informational' command
%ASA-5-111010：User 'enable_15'，running 'CLI' from IP 0.0.0.0，executed 'logging buffered informational'
Oct 20 2023 08:11:03: %ASA-5-111008: User 'enable_15' executed the 'logging timestamp' command
Oct 20 2023 08:11:03: %ASA-5-111010: User 'enable_15', running 'CLI' from IP 0.0.0.0, executed

'logging timestamp'

Oct 20 2023 08:11:08: %ASA-6-302010: 0 in use，3 most used

8. 配置允许通过 Telnet 远程登录

1）只允许主机 192.168.10.2 登录 ASA

cxyASA(config)#enable password cxypassword
cxyASA(config)#passwd cxypassword   //设置远程登录访问的登录密码
cxyASA(config)#telnet 192.168.10.2 255.255.255.255 inside
//开启 Telnet，只允许内部用户 192.168.10.2 登录
cxyASA(config)#username cxyuser password cxypassword privilege 15
//创建用户名及密码，用户权限级别为 15 级
cxyASA(config)#aaa authentication telnet console LOCAL //设置 telnet 的认证为本地认证
cxyASA(config)#

2）从路由器 R2 远程登录 ASA

cxyR2#telnet 192.168.10.100
Trying 192.168.10.100 ... Open
User Access Verification
Username：cxyuser
Password：***********
Your password was reset by the administrator.
Please change your password before proceeding.
Enter old password：***********
Enter new password：***********
Confirm new password：***********
User cxyuser logged into cxyASA
Logins over the last -2 days: 3. Last login: 09:16:58 UTC Oct 20 2023 from 192.168.10.2
Failed logins since the last login：0
Type help or '?' for a list of available_commands.
cxyASA>exit
Logoff
[Connection to 192.168.10.100 closed by foreign host]
cxyR2#

3）从路由器 R3 远程登录 ASA

cxyR3#telnet 192.168.10.100
Trying 192.168.10.100 ...
%Connection timed out；remote host not responding
cxyR3#

可见，可以从 R2 远程登录 ASA，但不能从 R3 远程登录 ASA。

9. 配置通过 SSH 安全远程登录

1）只允许管理员从 192.168.10.2 通过 SSH 安全远程登录 ASA

cxyASA(config)#ssh 192.168.10.2 255.255.255.255 inside
//开启 SSH 服务，只允许内部用户 192.168.10.2 登录
cxyASA(config)#aaa authentication ssh console LOCAL   //设置 SSH 的认证为本地认证
cxyASA(config)#crypto key generate rsa modulus 2048   //产生 RSA 密钥对
WARNING: You have a RSA keypair already defined named <Default-RSA-Key>.
Do you really want to replace them? [yes/no]: y
Keypair generation process begin. Please wait…
cxyASA(config)#ssh key-exchange hostkey rsa   //设置密钥交换为 rsa
cxyASA(config)#ssh cipher encryption custom "aes256-cbc:aes192-cbc:aes128-cbc"
//设置 SSH 采用的加密算法（列表，分隔符为":"）
cxyASA(config)#ssh cipher integrity custom "hmac-md5:hmac-sha1-96"
//设置 SSH 采用的完整性算法（列表，分隔符为":"）
cxyASA(config)#ssh key-exchange group dh-group14-sha1
//设置 SSH 采用的 Diffie-Hellman（DH）密钥交换模式
cxyASA(config)#ssh timeout 30   //设置 SSH 超时时长为 30s

2）从路由器 R2 安全远程登录 ASA

cxyR2#ssh -v 2 -l cxyuser -c aes128-cbc -m hmac-sha1-96 192.168.10.100
//可在 R2 中用"?"获得各开关的含义，这里-v 2 表示 SSH 的版本为 2
Password:
User cxy user logged into cxyASA
Logins over the last 1 days: 2. Last login: 09:28:33 UTC Oct 24 2023 from 192.168.10.2
Failed logins since the last login: 0. Last failed login: 09:28:25 UTC Oct 24 2023 from 192.168.10.2
Type help or '?' for a list of available commands.
cxyASA>exit
Logoff
[Connection to 192.168.10.100 closed by foreign host]

10. 配置 MPF

1）查看默认的 MPF

cxyASA#show running-config   //查看当前配置
……
policy-map global_policy   //在 ASA 中，已经设置了一个默认的 MPF
class inspection_default
inspect ip-options

```
   inspect netbios
   inspect rtsp
   inspect sunrpc
   inspect tftp
   inspect dns preset_dns_map
   inspect ftp
   inspect h323 h225
   inspect h323 ras
   inspect rsh
   inspect_esmtp
   inspect sqlnet
   inspect sip
!
service-policy global_policy global
……
```

可见，MPF 默认禁止 ICMP 数据包通过，故当内网主机 ping 外网主机时，外网主机的应答报文会被 ASA 的外部接口拒绝。也就是说，外网的 ICMP 报文是进入不了内网的。

2）在默认的 MPF global_policy 中允许 ICMP 报文进入，使内网 R2 可以 ping 通 R1

```
cxyASA(config)#policy-map global_policy
cxyASA(config-pmap)#class inspection_default
cxyASA(config-pmap-c)#inspect icmp
cxyASA(config-pmap-c)#exit
```

3）只查看默认的 MPF

```
cxyASA(config)#show running-config policy-map global_policy
// 注意比较此命令与前一个 show running-config 命令运行后显示的信息的异同
policy-map global_policy
 class inspection_default
  inspect ip-options
  inspect netbios
  inspect rtsp
  inspect sunrpc
  inspect tftp
  inspect dns preset_dns_map
  inspect ftp
  inspect h323 h225
  inspect h323 ras
```

```
  inspect rsh
  inspect_esmtp
  inspect sqlnet
  inspect sip
  inspect icmp
 !
 service-policy global_policy global
```

4）在 R2 上 ping R1

```
cxyR2#ping 202.100.0.1
Type escape sequence to abort.
Sending 5, 100-byte ICMP Echos to 202.100.0.1, timeout is 2 seconds:
!!!!!
Success rate is 100 percent(5/5), round-trip min/avg/max=56/60/64ms
```

可见，在 R2 上可以 ping 通 R1。

5）在 R1 上 ping R2

```
cxyR1#ping192.168.10.2
Type escape sequence to abort.
Sending 5, 100-byte ICMP Echos to 192.168.10.2, timeout is 2 seconds:
Success rate is percent (0/5)
```

可见，R1 是 ping 不通 R2 的，因为被 ASA 的外部接口拒绝了，可以通过日志查看被拒绝的信息。

6）查看日志，可看到被拒绝的信息

```
cxyASA(config)#show logging
Syslog logging: enabled
    Facility: 20
    Timestamp logging: enabled
    Timezone: disabled
    Hide Username logging: enabled
    Standby logging: disabled
    Debug-trace logging: disabled
    Console logging: disabled
    Monitor logging: disabled
    Buffer logging: level informational, 313 messages logged
    Trap logging: disabled
    Permit-hostdown logging: disabled
    History logging: disabled
    Device ID: disabled
```

Mail logging: disabled

ASDM logging: disabled

ded: IP address: 192.168.10.2, Uname: cxyuser

Oct 20 2023 13:07:28: %ASA-6-605005: Login permitted from 192.168.10.2/22747 to inside:192.168.10.100/telnet for user "cxyuser"

Oct 20 2023 13:08:07: %ASA-6-772005: REAUTH: user 'cxyuser' passed authentication

Oct 20 2023 13:09:18: %ASA-6-302014: Teardown TCP connection 73 for inside:192.168.10.2/22747 to identity:192.168.10.100/23 duration 0:02:09 bytes 544 TCP FINs from idy

Oct 20 2023 13:11:32: %ASA-6-110002: Failed to locate egress interface for TCP from DMZ:172.16.10.3/30224 to 192.168.10.100/23

Oct 20 2023 13:11:45: %ASA-6-302010: 0 in use, 3 most used

Oct 20 2023 13:16:16: %ASA-4-711004: Task ran for 115 msec, Process = ci/console, PC = e464fbd0, Call stack =

Oct 20 2023 13:16:16: %ASA-4-711004: Task ran for 115 msec, Process = ci/console, PC = e464fbd0, Call stack =   0x000055ebe464fbd0 0x000055ebe57f83d4 0x000055ebe57f8e

Oct 20 2023 13:21:46: %ASA-6-302010: 0 in use, 3 most used

Oct 20 2023 13:31:47: %ASA-6-302010: 0 in use, 3 most used

Oct 20 2023 13:34:14: %ASA-6-302013: Built inbound TCP connection 74 for inside:192.168.10.2/21913 (192.168.10.2/21913) to identity:192.168.10.100/22 (192.168.10.100/2)

Oct 20 2023 13:34:14: %ASA-6-315011: SSH session from 192.168.10.2 on interface inside for user "Unknown" disconnected by SSH server, reason: "Internal error" (0x00)

Oct 20 2023 13:34:14: %ASA-6-302014: Teardown TCP connection 74 for inside:192.168.10.2/21913 to identity:192.168.10.100/22 duration 0:00:00 bytes 259 TCP FINs from idy

Oct 20 2023 13:38:46: %ASA-6-302013: Built inbound TCP connection 75 for inside:192.168.10.2/27588 (192.168.10.2/27588) to identity:192.168.10.100/23 (192.168.10.100/2)

Oct 20 2023 13:38:46: %ASA-6-769007: UPDATE: Image version is 9.17(1)

Oct 20 2023 13:38:46: %ASA-4-769009: UPDATE: Image booted boot:/asa9171-smp-k8.bin is different from boot images

Oct 20 2023 13:38:50: %ASA-6-113015: AAA user authentication Rejected : reason = User was not found : local database : user = ***** : user IP = 192.168.10.2

Oct 20 2023 13:38:50: %ASA-6-611102: User authentication failed: IP address: 192.168.10.2, Uname: *****

Oct 20 2023 13:38:51: %ASA-6-302014: Teardown TCP connection 75 for inside:192.168.10.2/27588 to identity:192.168.10.100/23 duration 0:00:05 bytes 85 TCP FINs from idey

Oct 20 2023 13:40:03: %ASA-6-302013: Built inbound TCP connection 76 for inside:192.168.10.2/17259 (192.168.10.2/17259) to identity:192.168.10.100/22 (192.168.10.100/2)

Oct 20 2023 13:40:03: %ASA-6-315011: SSH session from 192.168.10.2 on interface inside for user "Unknown" disconnected by SSH server, reason: "Internal error" (0x00)

Oct 20 2023 13:40:03: %ASA-6-302014: Teardown TCP connection 76 for inside:192.168.10.2/

17259 to identity:192.168.10.100/22 duration 0:00:00 bytes 259 TCP FINs from idy

 Oct 20 2023 13:41:49: %ASA-6-302010: 0 in use, 3 most used

 Oct 20 2023 13:41:49: %ASA-6-302013: Built inbound TCP connection 77 for inside:192.168.10.2/47229 (192.168.10.2/47229) to identity:192.168.10.100/22 (192.168.10.100/2)

 Oct 20 2023 13:41:49: %ASA-6-315011: SSH session from 192.168.10.2 on interface inside for user "Unknown" disconnected by SSH server, reason: "Internal error" (0x00)

 Oct 20 2023 13:41:49: %ASA-6-302014: Teardown TCP connection 77 for inside:192.168.10.2/47229 to identity:192.168.10.100/22 duration 0:00:00 bytes 259 TCP FINs from idy

 Oct 20 2023 13:49:28: %ASA-3-106014: Deny inbound icmp src outside:202.100.0.1 dst inside:192.168.10.2 (type 8, code 0)

 Oct 20 2023 13:49:30: %ASA-3-106014: Deny inbound icmp src outside:202.100.0.1 dst inside:192.168.10.2 (type 8, code 0)

 Oct 20 2023 13:49:32: %ASA-3-106014: Deny inbound icmp src outside:202.100.0.1 dst inside:192.168.10.2 (type 8, code 0)

 Oct 20 2023 13:49:34: %ASA-3-106014: Deny inbound icmp src outside:202.100.0.1 dst inside:192.168.10.2 (type 8, code 0)

 Oct 20 2023 13:49:36: %ASA-3-106014: Deny inbound icmp src outside:202.100.0.1 dst inside:192.168.10.2 (type 8, code 0)

7）查看日志，只查看包含"icmp"的日志

cxyASA(config)#show logging | include icmp //只查看包含"icmp"的日志

 Oct 20 2023 13:49:28: %ASA-3-106014: Deny inbound icmp src outside:202.100.0.1 dst inside:192.168.10.2 (type 8, code 0)

 Oct 20 2023 13:49:30: %ASA-3-106014: Deny inbound icmp src outside:202.100.0.1 dst inside:192.168.10.2 (type 8, code 0)

 Oct 20 2023 13:49:32: %ASA-3-106014: Deny inbound icmp src outside:202.100.0.1 dst inside:192.168.10.2 (type 8, code 0)

 Oct 20 2023 13:49:34: %ASA-3-106014: Deny inbound icmp src outside:202.100.0.1 dst inside:192.168.10.2 (type 8, code 0)

 Oct 20 2023 13:49:36: %ASA-3-106014: Deny inbound icmp src outside:202.100.0.1 dst inside:192.168.10.2 (type 8, code 0)

11. 保存与导出 GNS3 配置文件

1）配置完成后，需要保存配置文件，否则断电或重启后配置将丢失

cxyASA(config)#write   //与 write memory 等效

或者：

cxyASA(config)#copy running-config startup-config   //与 write 等效

2）导出

（1）使用命令导出配置文件。

在 GNS3 中执行"Tools"→"Import/Export node configs"命令，按提示选择保存配置文件的文件夹，完成导出。但此方法有时不能保存全部设备的配置文件，如此项目中 ASA 的配置文件就不能导出，会出现提示信息"Config export is not supported by the following nodes: ASA"。可用下面"导出到 FTP 服务器"的方法导出 ASA 的配置文件。

（2）导出到 FTP 服务器。

①在物理机上创建 FTP 服务器。

在物理机中通过 IIS 创建 FTP 服务器，如图 13-2 所示。

图 13-2　通过 IIS 创建 FTP 服务器

②在 GNS3 中增加云 Cloud1。

在 GNS3 中增加云 Cloud1，将 ASA 的 G0/6 接口连接到 Cloud1 的 Eth1 接口（NAT 网络，虚拟机 NAT 网络的连接拓扑如图 13-3 所示）。

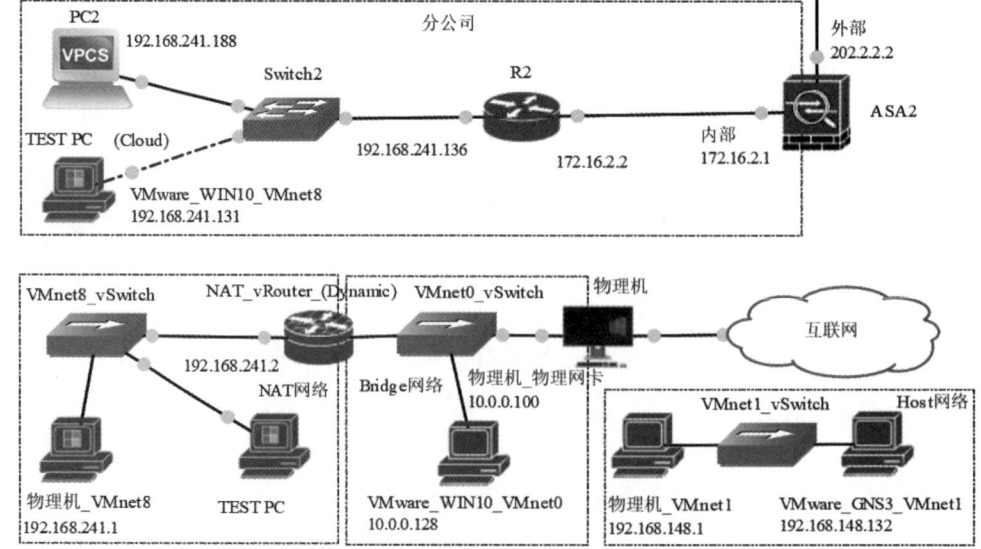

图 13-3　虚拟机 NAT 网络的连接拓扑

③配置 ASA。

```
cxyASA(config)#interface g0/6
cxyASA(config-if)#no shutdown
cxyASA(config-if)#nameif toftpsever
INFO: Security level for "toftpsever" set to 0 by default.
cxyASA(config-if)#security-level 100
cxyASA(config-if)#ip address dhcp
cxyASA(config)#show ip address
System IP Addresses：
```

Interface	Name	IP_address	Sub netmask	Method
GigabitEthernet 0/1	outside	202.100.0.100	255.255.255.0	CONFIG
GigabitEthernet 0/2	inside	192.168.10.100	255.255.255.0	CONFIG
GigabitEthernet 0/3	DMZ	172.16.10.100	255.255.255.0	CONFIG
GigabitEthernet 0/6	toftpsever	**192.168.241.144**	255.255.255.0	DHCP

可见，从 DHCP 服务器（此服务器在虚拟机的 NAT 网络中）获得的 IP 地址是 192.168.241.144，据此可得从 GNS3 网络到物理机的下一跳 IP 地址为 192.168.241.2（默认）。

```
cxyASA(config)#route toftpsever 10.0.0.0 255.0.0.0 192.168.241.2
cxyASA(config)#ping 10.0.0.100
Type escape sequence to abort.
Sending 5，100-byte ICMP Echos to 10.0.0.100，timeout is 2 seconds：
!!!!!
Success rate is 100 percent (5/5)，round-trip min/avg/max=1/2/10ms
```

④用 copy running-config 命令复制到 FTP 服务器。

```
cxyASA(config)#copy running-config ftp://10.0.0.100/ASA_running-config
Source filename [running-config] ?
Address or name of remote host [10.0.0.100] ?
Destination filename [ASA1_running-config] ?
Crypto checksum : 4e7c346 abf91e498 64880d3b 83d8edb5
12602 bytes copied in 0.280 secs
```
在上面三个 "?" 处，直接按回车键确认即可。

⑤打开 FTP 服务器，可以看到新下载的文件。

在物理机的 FTP 服务器中，可以看到刚刚从 ASA 中导出的名为 "ASA_running-config" 的配置文件，如图 13-4 所示。

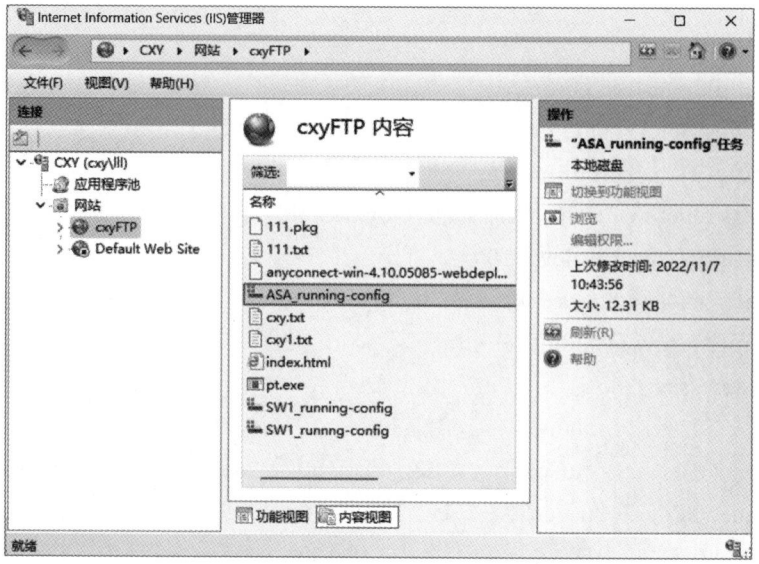

图 13-4　FTP 服务器中的 ASA 配置文件

第 14 章 ASA 防火墙的路由配置

14.1 组网需求

网络拓扑如图 14-1 所示，其中 ASA 为 Cisco ASA 系列防火墙，现要求在防火墙上实现以下功能。

（1）ASA 上配置默认路由，下一跳为 R1；
（2）ASA 上配置去往 33.33.33.33 的静态路由，下一跳为 R3；
（3）ASA 上配置 RIPv2，完成后通过 RIPv2 从 R3 学习到路由信息；
（4）ASA 上配置 OSPF，完成后通过 OSPF 从 R2 学习到路由信息；
（5）配置 ASA，将 RIPv2 导入 OSPF，使 R2 可以学习到 R3 的 Loopback0 接口的路由；
（6）配置 ASA，将默认路由导入 RIP，使 R2 可以通过 RIP 学习到默认路由。

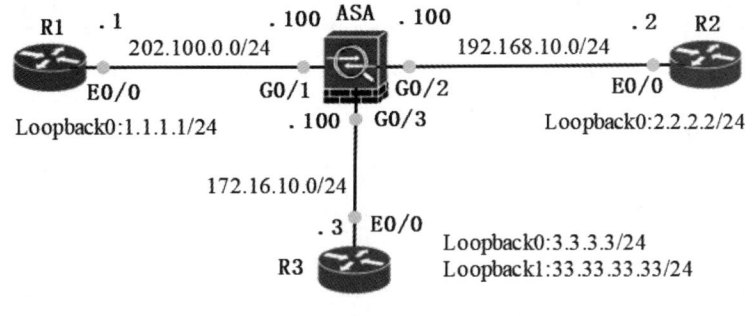

图 14-1 网络拓扑

14.2 仿真实验

14.2.1 仿真环境

仿真环境所需要的硬件、软件建议至少满足以下要求。

- 物理机的处理器主频为 1.6GHz；
- 内存 RAM 大小为 32.0GB；
- 物理机的操作系统为 Windows 10；
- 虚拟机为 VMware Workstation 16 Pro；

- 网络模拟器选用 GNS3 2.2.31；
- 防火墙为 Cisco ASAv9.17.1；
- 路由器为 Cisco IOL Router15.7。

14.2.2 实验设计

根据网络拓扑（图 14-1）完成网络设备的选型：防火墙选用 Cisco ASAv9.17.1，3 台路由器选用 Cisco IOL Router15.7。根据前面的组网需求，本实验设计如下。

1. IP 地址规划

网络有 3 个物理子网，网络地址分别为 202.100.0.0/24、192.168.10.0/24、172.16.10.0/24，此外还有 4 个环回 IP 地址。各设备接口的 IP 地址/掩码如表 14-1 所示。

表 14-1 各设备接口的 IP 地址/掩码

设备	接口	IP 地址/掩码
R1	E0/0	202.100.0.1/24
R1	Loopback0	1.1.1.1/24
R2	E0/0	192.168.10.2/24
R2	Loopback0	2.2.2.2/24
R3	E0/0	172.16.10.3/24
R3	Loopback0	3.3.3.3/24
R3	Loopback1	33.33.33.33/24
ASA	G0/1	202.100.0.100/24
ASA	G0/2	192.168.10.100/24
ASA	G0/3	172.16.10.100/24

2. 防火墙安全域设计

将 ASA 的接口 G0/1 定义为外部接口，安全级别采用默认值 0；接口 G0/2 定义为内部接口，安全级别采用默认值 100；接口 G0/3 定义为 DMZ 接口，安全级别设置为 50。

3. 基本路由设计

在防火墙 ASA 上先配置一条默认路由，下一跳为 R1（IP 地址为 202.100.0.1），然后再配置一条到达 33.33.33.33 的静态路由，下一跳为 R3（IP 地址为 172.16.10.3）。

在 ASA 和 R3 上启用 RIPv2，取消自动汇总；在 R3 的 172.16.0.0 和 3.0.0.0 上声明网络，在 ASA 的 172.16.0.0 上声明网络。

在 ASA 和 R2 上启用 OSPF，进程号为 1 且为主干区域，在 R2 的 192.168.10.0 和 2.2.2.0 上声明网络，在 ASA 的 192.168.10.0 上声明网络。

4. 路由导入设计

配置 ASA，将 RIPv2 得到的路由导入 OSPF，使 R2 可以学习到 R3 的 Loopback0 接口的路由。

配置 ASA，将默认路由导入 RIP，使 R3 可以通过 RIP 学习到默认路由。

14.2.3 操作步骤

下面给出本次实验的具体操作，一共包括 7 个步骤。

1. 按照图 14-1 所示的网络拓扑，在 GNS3 中新建一个空白的工程

根据网络拓扑设计网络，其中 R1~R3 选用 Cisco IOL Router15.7、ASA 选用 CiscoASAv9.17.1。配置完成后，在每个设备上均用"write"命令保存。

2. 配置路由器和防火墙接口的 IP 地址

1）根据设备接口的 IP 地址，配置路由器 R1 接口的 IP 地址

```
R1(config)#interface e0/0
R1(config-if)#ip address 202.100.0.1 255.255.255.0
R1(config-if)#no shutdown
R1(config-if)#exit
R1(config)#interface loopback0
R1(config-if)#ip address 1.1.1.1 255.255.255.0
R1(config-if)#exit
```

2）根据设备接口的 IP 地址，配置路由器 R2 接口的 IP 地址

```
R2(config)#interface e0/0
R2(config-if)#ip address 192.168.10.2 255.255.255.0
R2(config-if)#no shutdown
R2(config-if)#exit
R2(config)#interface loopback0
R2(config-if)#ip address 2.2.2.2 255.255.255.0
R2(config-if)#exit
```

3）根据设备接口的 IP 地址，配置路由器 R3 接口的 IP 地址

```
R3(config)#interface e0/0
R3(config-if)#ip address 172.16.10.3 255.255.255.0
R3(config-if)#no shutdown
R3(config-if)#exit
R3(config)#interface loopback0
R3(config-if)#ip address 3.3.3.3 255.255.255.0
R3(config-if)#exit
R3(config)#interface loopback1
R3(config-if)#ip address 33.33.33.33 255.255.255.0
R3(config-if)#exit
```

4）根据设备接口的 IP 地址，配置防火墙接口的 IP 地址

```
ASA(config)#hostname cxyASA
```

```
cxyASA(config)#interface g0/1
cxyASA(config-if)#nameif outside
INFO: Security level for "outside" set to 0 by default.
cxyASA(config-if)#ip address 202.100.0.100 255.255.255.0
cxyASA(config-if)#no shutdown
cxyASA(config-if)#exit
cxyASA(config)#interface g0/2
cxyASA(config-if)#nameif inside
INFO: Security level for "inside" set to 100 by default.
cxyASA(config-if)#ip address 192.168.10.100 255.255.255.0
cxyASA(config-if)#no shutdown
cxyASA(config-if)#exit
cxyASA(config)#interface g0/3
cxyASA(config-if)#nameif DMZ
INFO: Security level for "DMZ" set to 0 by default.
cxyASA(config-if)#security-level 50
cxyASA(config-if)#ip address 172.16.10.100 255.255.255.0
cxyASA(config-if)#no shutdown
cxyASA(config-if)#exit
```

5）测试直连网段连通性

```
cxyASA(config)#ping 202.100.0.1
Type escape sequence to abort.
Sending 5, 100-byte ICMP Echos to 202.100.0.1, timeout is 2 seconds:
!!!!!
Success rate is 100 percent(5/5), round-trip min/avg/max=10/10/10ms


cxyASA(config)#ping 192.168.10.2
Type escape sequence to abort.
Sending 5, 100-byte ICMP Echos to 192.168.10.2, timeout is 2 seconds:
!!!!!
Success rate is 100 percent(5/5), round-trip min/avg/max=1/10/20ms


cxyASA(config)#ping 172.16.10.3
Type escape sequence to_abort.
Sending 5, 100-byte ICMP Echos to 172.16.10.3, timeout is 2 seconds:
!!!!!
Success rate is 100percent(5/5), round-trip min/avg/max=1/8/10ms
```

3. 配置 ASA 静态路由

1）配置防火墙 ASA 的默认路由，下一跳为 R1

cxyASA(config)#route outside 0.0.0.0 0.0.0.0 202.100.0.1

2）配置去往 33.33.33.33 的静态路由，下一跳为 R3

cxyASA(config)#route DMZ 33.33.33.33 255.255.255.255 172.16.10.3

3）查看路由表

cxyASA(config)#show route
Codes：L-local，connected，S-static，R-RIP，M-mobile，B-BGP
   D_EIGRP，EX-EIGRP external，O-OSPF，IA-OSPF inter area
   N1-OSPF NSSA external type 1，N2-OSPF NSSA external type 2
   E1-OSPF external type 1，E2-OSPF external type 2，E-EGP
   i-IS-IS，L1-IS-IS level-l，L2-IS-IS level-2，ia-IS-IS inter area
   *-candidate default，U-per-user static route，o-ODR
   P-periodic downloaded static route

Gateway of last resort is 202.100.0.1 to network 0.0.0.0

```
S*      0.0.0.0 0.0.0.0 [1/0] via 202.100.0.1，outside
S       33.33.33.33 255.255.255.255 [1/0] via 172.16.10.3，DMZ
C       172.16.10.0 255.255.255.0 is directly connected，DMZ
L       172.16.10.100 255.255.255.255 is directly connected，DMZ
C       192.168.10.0 255.255.255.0 is directly connected，inside
L       192.168.10.100 255.255.255.255 is directly connected，inside
C       202.100.0.0 255.255.255.0 is directly connected，outside
L       202.100.0.100 255.255.255.255 is directly connected，outside
```

注意比较路由表在配置静态路由前后的变化。

4. 在 ASA 和 R3 上配置 RIP

1）在 R3 的 E0/0 和 Loopback0 接口上启用 RIPv2

R3(config)#router rip
R3(config-router)#version 2
R3(config-router)#no auto-summary
R3(config-router)#network 172.16.0.0
R3(config-router)#network 3.0.0.0
R3(config-router)#exit

2）在 ASA 的 G0/3 接口上启用 RIPv2

cxyASA(config)#router rip
cxyASA(config-router)#version 2

```
cxyASA(config-router)#no auto-summary
cxyASA(config-router)#network 172.16.0.0
cxyASA(config-router)#exit
```

3) 查看防火墙的路由表

```
cxyASA(config)#show route
Codes: L-local, connected, S-static, R-RIP, M-mobile, B-BGP
    D_EIGRP, EX-EIGRP external, O-OSPF, IA-OSPF inter area
    N1-OSPF NSSA external type 1, N2-OSPF NSSA external type 2
    E1-OSPF external type 1, E2-OSPF external type 2, E-EGP
    i-IS-IS, L1-IS-IS level-l, L2-IS-IS level-2, ia-IS-IS inter area
    *-candidate default, U-per-user static route, o-ODR
    P-periodic downloaded static route
Gateway of last resort is 202.100.0.1 to network 0.0.0.0
S*      0.0.0.0 0.0.0.0 [1/0] via 202.100.0.1, outside
R       3.3.3.0 255.255.255.0 [120/1] via 172.16.10.3, 00:00:07, DMZ
S       33.33.33.33 255.255.255.255 [1/0] via 172.16.10.3, DMZ
C       172.16.10.0 255.255.255.0 is directly connected, DMZ
L       172.16.10.100 255.255.255.255 is directly connected, DMZ
C       192.168.10.0 255.255.255.0 is directly connected, inside
L       192.168.10.100 255.255.255.255 is directly connected, inside
C       202.100.0.0 255.255.255.0 is directly connected, outside
L       202.100.0.100 255.255.255.255 is directly connected, outside
```

5. 在 ASA 和 R2 上配置 OSPF

1) **在 R2 的 E0/0 和 Loopback0 接口上启用 OSPF**

```
R2(config)#router ospf 1
R2(config-router)#router-id 22.22.22.22
R2(config-router)#network 192.168.10.0 0.0.0.255 area 0
R2(config-router)#network 2.2.2.0 0.0.0.255 area 0
R2(config-router)#exit
```

2) **在 ASA 的 G0/2 接口上启用 OSPF**

```
cxyASA(config)#router ospf 1
cxyASA(config-router)#router-id 2.2.2.2
cxyASA(config-router)#network 192.168.10.0 255.255.255.0 area 0
//ASA 不支持通配符（OSPF 中不要使用反子网掩码）
cxyASA(config-router)#exit
```

3）查看防火墙的路由表

```
cxyASA(config)#show route
Codes：L-local，connected，S-static，R-RIP，M-mobile，B-BGP
    D_EIGRP，EX-EIGRP external，O-OSPF，IA-OSPF inter area
    N1-OSPF NSSA external type 1，N2-OSPF NSSA external type 2
    E1-OSPF external type 1，E2-OSPF external type 2，E-EGP
    i-IS-IS，L1-IS-IS level-l，L2-IS-IS level-2，ia-IS-IS inter area
    *-candidate default，U-per-user static route，o-ODR
    P-periodic downloaded static route
Gateway of last resort is 202.100.0.1 to network 0.0.0.0
S*      0.0.0.00.0.0.0 [1/0] via 202.100.0.1，outside
O       2.2.2.2 255.255.255.255 [110/11] via 192.168.10.2，00:00:54，inside
R       3.3.3.0 255.255.255.0 [120/1] via 172.16.10.3, 00:00:07, DMZ
S       33.33.33.33 255.255.255.255 [1/0] via 172.16.10.3，DMZ
C       172.16.10.0 255.255.255.0 is directly connected，DMZ
L       172.16.10.100 255.255.255.255 is directly connected，DMZ
C       192.168.10.0 255.255.255.0 is directly connected，inside
L       192.168.10.100 255.255.255.255 is directly connected，inside
C       202.100.0.0 255.255.255.0 is directly connected，outside
L       202.100.0.100 255.255.255.255 is directly connected，outside
```

6. 动态路由间的导入

1）配置路由导入前，查看 R2 的路由表

```
R2#show ip route
Codes：L-local，connected，S-static，R-RIP，M-mobile，B-BGP
    D_EIGRP，EX-EIGRP external，O-OSPF，IA-OSPF inter area
    N1-OSPF NSSA external type 1，N2-OSPF NSSA external type 2
    E1-OSPF external type 1，E2-OSPF external type 2，E-EGP
    i-IS-IS，L1-IS-IS level-l，L2-IS-IS level-2，ia-IS-IS inter area
    *-candidate default，U-per-user static route，o-ODR
    P-periodic downloaded static route
Gateway of last resort is 192.168.10.100 to network 0.0.0.0
S*      0.0.0.0/0[1/0] via 192.168.10.100
        2.0.0.0/8 is variably subnetted，2 subnets，2masks
C       2.2.2.0/24 is directly connected，Loopback 0
L       2.2.2.2/32 is directly connected，Loopback 0
        192.168.10.0/24 is variably subnetted，2 subnets，2masks
C       192.168.10.0/24 is directly connected，Ethernet 0/0
L       192.168.10.2/32 is directly connected，Ethernet 0/0
```

2）配置 ASA，将 RIPv2 得到的路由导入 OSPF 中，使 R2 可以学习到 R3 的 Loopback0 接口的路由

```
cxyASA(config)#router ospf 1
cxyASA(config-router)#redistribute rip subnets
cxyASA(config-router)#exit
```

3）查看 R2 的路由表

```
R2#show ip route
Codes: L-local, connected, S-static, R-RIP, M-mobile, B-BGP
    D_EIGRP, EX-EIGRP external, O-OSPF, IA-OSPF inter area
    N1-OSPF NSSA external type 1, N2-OSPF NSSA external type 2
    E1-OSPF external type 1, E2-OSPF external type 2, E-EGP
    i-IS-IS, L1-IS-IS level-l, L2-IS-IS level-2, ia-IS-IS inter area
    *-candidate default, U-per-user static route, o-ODR
    P-periodic downloaded static route
Gateway of last resort is 192.168.10.100 to network 0.0.0.0
S*        0.0.0.0/0[1/0] via 192.168.10.100
          2.0.0.0/8 is variably subnetted, 2 subnets, 2 masks
C         2.2.2.0/24 is directly connected, Loopback 0
L         2.2.2.2/32 is directly connected, Loopback 0
          3.0.0.0/24 is subnetted, 1 subnets
O E2      3.3.3.0 [110/20] via 192.168.10.100, 00:00:19, Ethernet 0/0
          172.16.0.0/24 is subnetted, 1 subnets
O E2      172.16.10.0 [110/20] via 192.168.10.100, 00:00:19, Ethernet 0/0
          192.168.10.0/24 is variably subnetted, 2 subnets, 2 masks
C         192.168.10.0/24 is directly connected, Ethernet 0/0
L         192.168.10.2/32 is directly connected, Ethernet 0/0
```

7. 默认路由导入 RIP

1）配置路由导入前，查看 R3 的路由表

```
R3#show ip route
Codes: L-local, connected, S-static, R-RIP, M-mobile, B-BGP
    D_EIGRP, EX-EIGRP external, O-OSPF, IA-OSPF inter area
    N1-OSPF NSSA external type 1, N2-OSPF NSSA external type 2
    E1-OSPF external type 1, E2-OSPF external type 2, E-EGP
    i-IS-IS, L1-IS-IS level-l, L2-IS-IS level-2, ia-IS-IS inter area
    *-candidate default, U-per-user static route, o-ODR
    P-periodic downloaded static route
```

Gateway of last resort is not set
    3.0.0.0/8 is variably subnetted, 2 subnets, 2 masks
C   3.3.3.0/24 is directly connected，Loopback 0
L   3.3.3.3/32 is directly connected，Loopback 0
    33.0.0.0/8 is variably subnetted, 2 subnets, 2 masks
C   33.33.33.0/24 is directly connected, Loopback 1
L   33.33.33.33/32 is directly connected, Loopback 1
    172.16.0.0/16 is variably subnetted, 2 subnets, 2 masks
C   172.16.10.0/24 is directly connected, Ethernet 0/0
L   172.16.10.3/32 is directly connected, Ethernet 0/0

2）配置 ASA，将默认路由导入 RIP，使 R3 可以通过 RIP 学习到默认路由

cxyASA(config)#router rip
cxyASA(config-router)#redistribute static metric 5
cxyASA(config-router)#exit

3）查看 R3 的路由表

R3#show ip route
des：L-local，connected，S-static，R-RIP，M-mobile，B-BGP
    D_EIGRP，EX-EIGRP external，O-OSPF，IA-OSPF inter area
    N1-OSPF NSSA external type 1，N2-OSPF NSSA external type 2
    E1-OSPF external type 1，E2-OSPF external type 2，E-EGP
    i-IS-IS，L1-IS-IS level-l，L2-IS-IS level-2，ia-IS-IS inter area
    *-candidate default，U-per-user static route，o-ODR
    P-periodic downloaded static route
Gateway of last resort is **172.16.10.100 to network 0.0.0.0**
R* 0.0.0.0/0 [120/5] via 172.16.10.100，00:00:05, Ethernet 0/0
    3.0.0.0/8 is variably subnetted, 2 subnets, 2 masks
C   3.3.3.0/24 is directly connected，Loopback 0
L   3.3.3.3/32 is directly connected，Loopback 0
    33.0.0.0/8 is variably subnetted, 2 subnets, 2 masks
C   33.33.33.0/24 is directly connected, Loopback 1
L   33.33.33.33/32 is directly connected, Loopback 1
    172.16.0.0/16 is variably subnetted, 2 subnets, 2 masks
C   172.16.10.0/24 is directly connected, Ethernet 0/0
L   172.16.10.3/32 is directly connected, Ethernet 0/0

第 15 章
ASA 防火墙的 NAT 和 ACL 配置

15.1 组网需求

网络拓扑如图 15-1 所示，其中 ASA 为 Cisco ASA 系列防火墙，现要求在防火墙上实现以下功能。

（1）配置 PAT（Port Address Translation，端口地址转换），实现内部区域内主机访问互联网。

（2）配置静态 NAT（Network Address Translation，网络地址转换），实现 DMZ 区域内主机的 IP 地址 172.16.2.22 转换为 202.100.0.222。

（3）配置 Identity NAT，实现 172.16.110 访问外部区域时，不进行地址转换。

（4）配置 ACL，允许 DMZ 区域内的 ICMP 和 Telnet 报文流向内部区域。

（5）配置 ACL，实现内部区域的 Server0 只能访问本区域内的主机。不允许去往其他区域。

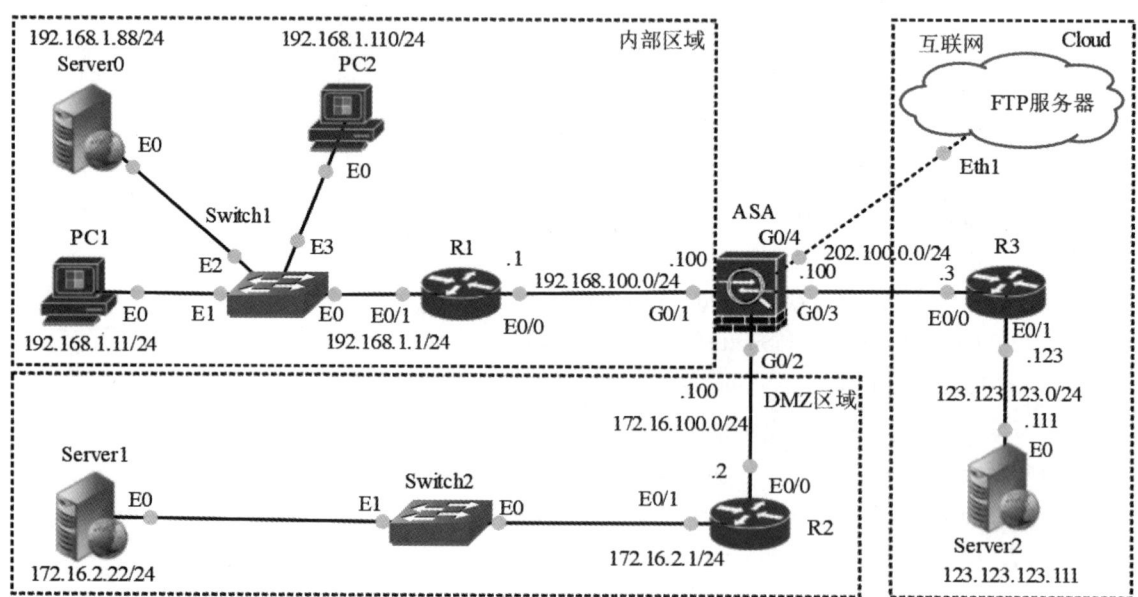

图 15-1 网络拓扑

15.2 仿真实验

15.2.1 仿真环境

仿真环境所需要的硬件、软件建议至少满足以下要求。

- 物理机的处理器主频为 1.6GHz；
- 内存 RAM 大小为 32.0GB；
- 物理机的操作系统为 Windows 10；
- 虚拟机为 VMware Workstation 16 Pro；
- 网络模拟器选用 GNS3 2.2.31；
- 防火墙为 Cisco ASAv9.17.1；
- 交换机为 Ethernet switch；
- 路由器为 Cisco IOL Router15.7；
- PC1 和 PC2 为 Virtual PC Simulator, version 0.8.2。

15.2.2 实验设计

根据网络拓扑（图 15-1）完成网络设备的选型：防火墙选用 Cisco ASAv9.17.1，3 台路由器选用 Cisco IOL Router15.7。根据前面的组网需求，本实验设计如下。

1. IP 地址规划

网络有 6 个物理子网，网络地址分别为 192.168.100.0/24、192.168.1.0/24、172.16.100.0/24、172.16.2.0/24、202.100.0.0/24、123.123.123.0/24。各设备接口的 IP 地址/掩码如表 15-1 所示。

表 15-1 各设备接口的 IP 地址/掩码

设备	接口	IP 地址/掩码
R1	E0/0	192.168.100.1/24
R1	E0/1	192.168.1.1/24
R2	E0/0	172.16.100.2/24
R2	E0/1	172.16.2.1/24
R3	E0/0	202.100.0.3/24
R3	E0/1	123.123.123.123/24
ASA	G0/1	192.168.100.100/24
ASA	G0/2	172.16.100.100/24
ASA	G0/3	202.100.0.100/24
ASA	G0/4	—
PC1 和 PC2、Server0～Server2	E0	见图 15-1

2. 路由设计

在路由器 R1 上配置一条下一跳地址为 192.168.100.100 的默认路由，在路由器 R2 上配置一条下一跳地址为 172.16.100.100 的默认路由。

在防火墙 ASA 上先配置一条下一跳地址为 202.100.0.3 的默认路由，然后定义一条到达 192.168.1.0/24 网络的静态路由（下一跳 IP 地址为 192.168.100.1），最后增加一条到达

172.16.2.0/24 网络的静态路由（下一跳 IP 地址为 172.16.100.2）。

3. 防火墙设计

首先，定义安全域：将 ASA 的接口 G0/3 定义为外部接口，安全级别采用默认值 0；将接口 G0/1 定义为内部接口，安全级别采用默认值 100；将接口 G0/2 定义为 DMZ 接口，安全级别设置为 50。

然后，在 ASA 上完成下列操作：配置 MPF 允许 ICMP 报文，实现等级高的区域可以 ping 通等级低的区域；配置 PAT，实现内部区域内的主机可以访问互联网；配置静态地址转换，实现 DMZ 区域主机 172.16.2.22 转换为 202.100.0.222；配置 Identity NAT，实现特定 IP 地址不要进行地址转换。

由于防火墙默认的安全策略是安全级别低的区域不能访问安全级别高的区域，因此需要配置 ACL，允许 DMZ 区域内主机的 ICMP 和 Telnet 报文流向内部区域；配置扩展 ACL，实现内部区域内的 Server0（192.168.1.88）只能访问本区域，而不能访问其他区域。

15.2.3　操作步骤

下面给出本次实验的具体操作，一共包括 9 个步骤。

1. 按照图 15-1 所示的网络拓扑，在 GNS3 中新建一个空白的工程

按照网络拓扑设计网络，其中 R1～R3 选用 Cisco IOL Router15.7，ASA 选用 Cisco ASAv9.17.1，PC1、PC2 及 Server0～Server2 均选用 VPCS（在 GNS 网络拓扑中右击待更改图标的设备，在弹出的快捷菜单中选择"change symbol"命令，在打开的对话框中单击"custom symbols"项即可找到合适的图标）。

FTP 服务器通过 Cloud 的 Eth1 接口（对应虚拟网卡的 VMware Network Adapter VMnet8 网络，IP 地址为 192.168.241.1/24）接入 GNS3 网络中。

2. 配置路由器和防火墙接口的 IP 地址

1）根据设备接口的 IP 地址，配置路由器 R1 接口的 IP 地址

```
R1(config)#hostname cxyR1          // 可将 cxy 改为你的名字拼音首字母
cxyR1(config)#interface e0/0
cxyR1(config-if)#ip address 192.168.100.1 255.255.255.0
cxyR1(config-if)#no shutdown
cxyR1(config-if)#exit
cxyR1(config)#interface e0/1
cxyR1(config-if)#ip address 192.168.1.1 255.255.255.0
cxyR1(config-if)#no shutdown
cxyR1(config-if)#exit
```

2）根据设备接口的 IP 地址，配置路由器 R2 接口的 IP 地址

```
R2(config)#hostname cxyR2          // 可将 cxy 改为你的名字拼音首字母
cxyR2(config)#interface e0/0
```

```
cxyR2(config-if)#ip address 172.16.100.2 255.255.255.0
cxyR2(config-if)#no shutdown
cxyR2(config-if)#exit
cxyR2(config)#interface e0/1
cxyR2(config-if)#ip address 172.16.2.1 255.255.255.0
cxyR2(config-if)#no shutdown
cxyR2(config-if)#exit
```

3）根据设备接口的 IP 地址，配置路由器 R3 接口的 IP 地址

```
R3(config)#hostname cxyR3          // 可将 cxy 改为你的名字拼音首字母
cxyR3(config)#interface e0/0
cxyR3(config-if)#ip address 202.100.0.3 255.255.255.0
cxyR3(config-if)#no shutdown
cxyR3(config-if)#exit
cxyR3(config)#interface e0/1
cxyR3(config-if)#ip address 123.123.123.123 255.255.255.0
cxyR3(config-if)#no shutdown
cxyR3(config-if)#exit
```

4）根据设备接口的 IP 地址，配置防火墙接口的 IP 地址

```
CISCOASA (config)#hostname cxyASA       // 可将 cxy 改为你的名字拼音首字母
cxyASA(config)#interface g0/1
cxyASA(config-if)#nameif inside
INFO: Security level for "inside" set to 100 by default.
cxyASA(config-if)#ip address 192.168.100.100 255.255.255.0
cxyASA(config-if)#no shutdown
cxyASA(config-if)#exit
cxyASA(config)#interface g0/2
cxyASA(config-if)#nameif DMZ
INFO: Security level for "DMZ" set to 0 by default.
cxyASA(config-if)#security-level 50
cxyASA(config-if)#ip address 172.16.100.100 255.255.255.0
cxyASA(config-if)#no shutdown
cxyASA(config-if)#exit
cxyASA(config)#interface g0/3
cxyASA(config-if)#nameif outside
INFO: Security level for "outside" set to 0 by default.
cxyASA(config-if)#ip address 202.100.0.100 255.255.255.0
cxyASA(config-if)#no shutdown
cxyASA(config-if)#exit
```

5）测试直连网络的连通性

cxyASA(config)#ping 192.168.100.1
Type escape sequence to abort.
Sending 5，100-byte ICMP Echos to 192.168.100.1，timeout is 2 seconds：
!!!!!
Success rate is 100 percent (5/5)，round-trip min/avg/max=1/1/1ms

cxyASA(config)#ping 172.16.100.2
Type escape sequence to abort.
Sending 5，100-byte ICMP Echos to 172.16.100.2，timeout is 2 seconds：
!!!!!
Success rate is 100 percent (5/5)，round-trip min/avg/max=1/1/1ms

cxyASA(config)#ping 202.100.0.3
Type escape sequence to abort.
Sending 5，100-byte ICMP Echos to 202.100.0.3，timeout is 2 seconds：
!!!!!
Success rate is 100 percent (5/5)，round-trip min/avg/max=1/1/1ms

3. 配置静态路由，实现内部区域、DMZ 区域网络层互通（外部区域除外，R3 不用添加路由）

cxyR1(config)#ip route 0.0.0.0 0.0.0.0 192.168.100.100   //默认路由到 ASA
cxyR2(config)#ip route 0.0.0.0 0.0.0.0 172.16.100.100   //默认路由到 ASA
cxyASA(config)#route inside 192.168.1.0 255.255.255.0 192.168.100.1
cxyASA(config)#route outside 0.0.0.0 0.0.0.0 202.100.0.3
cxyASA(config)#route DMZ 172.16.2.0 255.255.255.0 172.16.100.2

4. 配置 MPF 允许 ICMP 报文，实现等级高的区域可以 ping 通等级低的区域

cxyASA(config)#policy-map global_policy
cxyASA(config-pmap)#class inspection_default
cxyASA(config-pmap-c)#inspect icmp
cxyASA(config-pmap-c)#exit
cxyASA(config-pmap)#exit

5. 配置 PAT，使内部区域内的主机可以访问互联网
1）多对一 PAT 配置

cxyASA (config)#object network my-inside-net   //创建名为 my-inside-net 的网络对象
cxyASA (config-network-object)#subnet 192.168.1.0 255.255.255.0
//向 my-inside-net 网络对象添加子网 192.168.1.0/24

cxyASA (config-network-object)#nat (inside,outside) dynamic interface
//允许源地址是子网 192.168.1.0/24 内主机的 IP 地址转换为外部接口的公网 IP 地址
cxyASA(config)#

2）多对一 PAT 调试

（1）从内网主机 PC1 ping 互联网上的服务器 Server2。

PC1>ping 123.123.123.111
84 bytes from 123.1 23.123.111 icmp_seq=1ttl= 62 time =2.278ms
84 bytes from 123.1 23.123.111 icmp_seq=2ttl= 62 time =1.753ms
84 bytes from 123.1 23.123.111 icmp_seq=3ttl= 62 time =1.108ms
84 bytes from 123.123.123.111 icmp_seq= 4ttl= 62 time =1.164ms
84 bytes from 123.1 23.123.111 icmp_seq=5ttl= 62 time=1.114ms

（2）查看 NAT 表。

cxyASA(config)#show xlate
6 in use, 11 most used
Flags：D-DNS, e-extended, I-identity, i-dynamic, r-portmap, s-static, T-twice, N-net-to-net
NAT from inside: 192.168.100.1 to outside: 202.100.0.188
　　Flags s idle 1:13:40 timeout 0:00:00

ICMP PAT from inside:192.168.1.11/50337 to outside 202.100.0.100/50337 flags ri idle 0:00:08 timeout 0:00:30
　　ICMP PAT from inside:192.168.1.11/50081 to outside:202.100.0.100/50081 flags ri idle 0:00:09 timeout 0:00:30
　　ICMP PAT from inside:92.168.1.11/49825 to outside:202.100.0.100/50081 flags ri idle 0:00:10 timeout 0:00:30
　　ICMP PAT from inside:92.168.1.11/49569 to outside:202.100.0.100/49569 flags ri idle 0:00:11 timeout 0:00:30
　　ICMP PAT from inside:92.168.1.11/49313 to outside:202.100.0.100/49313 flags ri idle 0:00:12 timeout 0:00:30

（3）从内网服务器 Server0 上 ping 互联网上的服务器 Server2。

Server0>ping 123.123.123.111
84 bytes from 123.123.123.111 icmp_seq=1 ttl= 62 time =3.945 mS
84 bytes from 123.123.123.111 icmp_seq=2 ttl= 62 time =1.048 ms
84 bytes from 123.123.123.111 icmp_seq=3 ttl= 62 time =1.112 ms
84 bytes from 123.123.123.111 icmp_seq=4 ttl= 62 time =1.636 ms
84 bytes from 123.123.123.111 icmp_seq=5 ttl= 62 time =1.336 ms

（4）查看 NAT 表。

```
cxyASA(config)#show xlate
6 in use, 11 most used
Flags：D-DNS, e-extended, I-identity, i-dynamic, r-portmap, s-static, T-twice, N-net-to-net
NAT from inside: 192.168.100.1 to outside: 202.100.0.188
    Flags s idle 1:30:08 timeout 0:00:00

ICMP PAT from inside:192.168.1.88/37029 to outside 202.100.0.100/37029 flags ri idle 0:00:23 timeout 0:00:30
ICMP PAT from inside:192.168.1.88/36773 to outside 202.100.0.100/36773 flags ri idle 0:00:24 timeout 0:00:30
ICMP PAT from inside:192.168.1.88/36517 to outside 202.100.0.100/36517 flags ri idle 0:00:25 timeout 0:00:30
ICMP PAT from inside:192.168.1.88/36261 to outside 202.100.0.100/36261 flags ri idle 0:00:26 timeout 0:00:30
ICMP PAT from inside:192.168.1.88/36005 to outside 202.100.0.100/36005 flags ri idle 0:00:27 timeout 0:00:30
```

对比从 PC1 和 Server0 上分别 ping Server2，防火墙的 NAT 表中转换后的源 IP 地址有何不同，请分析其原因。

（5）从 Internet 上的 Server2 ping 内网主机 PC1。

```
Server2> ping 192.168.1.11
    *123.123.123.123 icmp_seq=1 ttl=255 time=0.168 ms (ICMP type:3, code:1, Destination host unreachable)
    *123.123.123.123 icmp_seq=2 ttl= 255 time=0.239 ms (ICMP type:3, code:1, Destination host unreachable)
    *123.123.123.123 icmp_seq=3 ttl= 255 time=0.409 ms (ICMP type:3, code:1, Destination host unreachable)
    *123.123.123.123 icmp_seq=4 ttl= 255 time=0.310 ms (ICMP type:3, code:1, Destination host unreachable)
    *123.123.123.123 icmp_seq=5 ttl= 255 time=0.238 ms (ICMP type:3, code:1, Destination host unreachable)
```

可见，从互联网上的服务器不能 ping 通内网中的主机，请分析其原因。

6. 配置静态 NAT

1）配置静态 NAT，实现 DMZ 区域内主机的 IP 地址 172.16.2.22 转换为 202.100.0.222

```
cxyASA (config)#object network my-DMZ-Server      //创建网络对象
cxyASA (config-network-object)#host 172.16.2.22
//向网络对象 my-DMZ-Server 添加 IP 地址为 172.16.2.22 的主机
cxyASA (config-network-object)#nat (DMZ, outside) static 202.100.0.222
```

//允许主机的 IP 地址 172.16.2.22 的静态转换为公网 IP 地址 202.100.0.222

2）查看 NAT 表

cxyASA(config)#show xlate
1 in use，17 most used
Flags：D-DNS, e-extended, I-identity, i_dynamic, r-portmap，
　　　S-static, T-twice, N-net-to-net
NAT from DMZ: 172.16.2.22 to outside: 202.100.0.222
Flags s idle 0:02:44 timeout 0:00:00

3）静态 NAT 调试

（1）从 DMZ 区域的 Server1 上 ping 互联网上的 Server2。

Server1>ping 123.123.123.111
84 bytes from 123.123.123.111 icmp_seq=1 ttl=62 time=5.077 ms
84 bytes from 123.123.123.111 icmp_seq=2 ttl=62 time=1.749 ms
84 bytes from 123.123.123.111 icmp_seq=3 ttl=62 time=1.660 ms
84 bytes from 123.123.123.111 icmp_seq=4 ttl=62 time=1.398 ms
84 bytes from 123.123.123.111 icmp_seq=5 ttl=62 time=1.852 ms

可见，可以从 DMZ 区域中的主机访问互联网上的服务器。

（2）从互联网上的 Server2 ping DMZ 区域中的 Server1 的对外公网 IP 地址 202.100.0.222。

Server2> ping 202.100.0.222
202.100.0.222 icmp_seq=1 timeout
202.100.0.222 icmp_seq=2 timeout
202.100.0.222 icmp_seq=3timeout
202.100.0.222 icmp_seq=4 timeout
202.100.0.222 icmp_seq=5 timeout

可见，从互联网（外部区域）不能访问 DMZ 区域中的服务器，解释其原因。

（3）配置 ACL，允许外部区域访问 DMZ 区域。

cxyASA(config)#access-list DMZ-to-ISP extended permit icmp any object my-DMZ-Server
//允许任意主机向 DMZ 服务器发送 ICMP 报文
cxyASA(config)# access-group DMZ-to-ISP in interface outside
//在外部接口上应用名为 DMZ-to-ISP 的 ACL

（4）再次从互联网的 Server2 ping DMZ 区域中的 Server1 的对外公网 IP 地址 202.100.0.222。

Server2> ping 202.100.0.222
84 bytes from 202 .100.0.222 icmp_seq=1 ttl=62 time=3.503 ms
84 bytes from 202 .100.0.222 icmp_seq=2 ttl=62 time=1.907 ms
84 bytes from 202 .100.0.222 icmp_seq=3 ttl=62 time=2.042 ms

```
84 bytes from 202.100.0.222 icmp_seq=4 ttl=62 time=1.594 ms
84 bytes from 202.100.0.222 icmp_seq=5 ttl=62 time=2.609 ms
```

(5) 打开 R3 的 ICMP 调试开关,以便观察到 NAT 的效果。

```
cxyR3#debug ip icmp    //打开 ICMP 调试开关
ICMP packet debugging is on
```

(6) 从 DMZ 区域中的 Server1 ping 互联网上的路由器 R3 的 IP 地址 123.123.123.123。

```
Server1>ping 123.123.123.123
84 bytes from 123.123.123.123 icmp_seq=1 ttl=254 time=2.830 ms
84 bytes from 123.123.123.123 icmp_seq=2 ttl=254 time=1.755 ms
84 bytes from 123.123.123.123 icmp_seq=3 ttl=254 time=1.841 ms
84 bytes from 123.123.123.123 icmp_seq=4 ttl=254 time=1.785 ms
84 bytes from 123.123.123.123 icmp_seq=5 ttl=254 time=1.443 ms
```

(7) 在 R3 上观察 NAT 的效果。

```
cxyR3#
    *Oct 25 14:29:36.651: ICMP: echo reply sent, src 123.123.123.123, dst 202.100.0.222, topology BASE, dscp 0 topoid 0
    *Oct 25 14:29:36.651: ICMP: echo reply sent, src 123.123.123.123, dst 202.100.0.222, topology BASE, dscp 0 topoid 0
cxyR3#
    *Oct 25 14:29:36.651: ICMP: echo reply sent, src 123.123.123.123, dst 202.100.0.222, topology BASE, dscp 0 topoid 0
    *Oct 25 14:29:36.651: ICMP: echo reply sent, src 123.123.123.123, dst 202.100.0.222, topology BASE, dscp 0 topoid 0
```

7. 配置 Identity NAT

1)分析配置 Identity NAT 之前主机 PC2 访问外部区域的过程

(1) 在主机 PC2(192.168.1.110)上 ping 互联网上的路由器 R3。

```
PC2> ping 123.123.123.123
84 bytes from 123.123.123.123 icmp_seq=1 ttl= 254 time =2.769 ms
84 bytes from 123.123.123.123 icmp_seq=2 ttl= 254 time =1.133 ms
84 bytes from 123.123.123.123 icmp_seq=3 ttl= 254 time =1.354 ms
84 bytes from 123.123.123.123 icmp_seq=4 ttl= 254 time =1.228 ms
84 bytes from 123.123.123.123 icmp_seq=5 ttl= 254 time =1.105 ms
```

(2) 在路由器 R3 上查看 NAT 结果。

```
cxyR3#
    *Oct 25 15:42:06.181: ICMP: echo reply sent, src 123.123.123.123, dst 202.100.0.100, topology
```

BASE, dscp 0 topoid 0

  *Oct 25 15:42:07.184: ICMP: echo reply sent, src 123.123.123.123, dst **202.100.0.100**,topology BASE, dscp 0 topoid 0

  cxyR3#

  *Oct 25 15:42:08.185: ICMP: echo reply sent, src 123.123.123.123, dst **202.100.0.100**,topology BASE, dscp 0 topoid 0

  *Oct 25 15:42:09.188: ICMP: echo reply sent, src 123.123.123.123, dst **202.100.0.100**,topology BASE, dscp 0 topoid 0

可见，此时 PC2（192.168.1.110）ping R3（123.123.123.123），R3 接收到的是被转换后的 IP 地址（202.100.0.100）。

2）配置 Identity NAT，实现特定 IP 地址不要进行地址转换

（1）配置 Identity NAT，实现内部区域的主机 PC2 访问外部区域时，不要进行地址转换。

  cxyASA (config)#object network my-host-identity //创建网络对象
  cxyASA (config-network-object)#host 192.168.1.110 //向网络对象添加主机 PC2
  cxyASA (config-network-object)#nat (inside,outside) static 192.168.1.110
  //允许源 IP 地址是 192.168.1.110 的主机执行 Identity NAT，即不转换源 IP 地址

（2）调试 Identity NAT，在 PC2 上 ping 互联网上的路由器 R3。

  PC2> ping 123.123.123.123
  123.123.123.123 icmp_seq= 1 timeout
  123.123.123.123 icmp_seq= 2 timeout
  123.123.123.123 icmp_seq= 3 timeout
  123.123.123.123 icmp_seq= 4 timeout
  123.123.123.123 icmp_seq= 5 timeout

可见，此时 PC2 不能 ping 通路由器 R3（123.123.123.123）。

（3）在 R3 中，发现 192.168.1.110 没有被地址转换。

  cxyR3#

  *Oct 25 15:45:42.073: ICMP: echo reply sent, src 123.123.123.123, dst **192.168.1.110**,topology BASE, dscp 0 topoid 0

  cxyR3#

  *Oct 25 15:45:44.072: ICMP: echo reply sent, src 123.123.123.123, dst **192.168.1.110**,topology BASE, dscp 0 topoid 0

  cxyR3#

  *Oct 25 15:45:46.074: ICMP: echo reply sent, src 123.123.123.123, dst **192.168.1.110**,topology BASE, dscp 0 topoid 0

（4）查看 R3 的路由表。

  cxyR3#show ip route

```
Codes：L-local，C-connected，S-static，R-RIP，M-mobile，B-BGP
       D-EIGRP，EX-EIGRP external，O-OSPF，I A-OSPF inter area
       N1-OSPF NSSA external type 1，N2-OSPF NSSA external type 2
       E1-OSPF external type1，E2-OSPF external type 2
       i-IS-IS，su-IS-IS summary，L1-IS-IS level-1，L2-IS-IS level-2
       ia-IS-IS inter area，*- candidate default，U-per-user static route
       o-ODR，P-periodic downloaded static route，H-NHRP，l-LISP
       a-application route
       +-replicated route，%-next hop override，p-overrides from PfR
Gateway of last resort is not set
        123.0.0.0/8 is variably subnetted，2 subnets，2 masks
C       123.123.123.0/24 is directly connected，Ethernet 0/1
L       123.123.123.123/32 is directly connected，Ethernet 0/1
        202.100.0.0/24 is variably subnetted，2 subnets，2 masks
C       202.100.0.0/24 is directly connected，Ethernet 0/0
L       202.100.0.3/32 is directly connected，Ethernet 0/0
```

可见，R3 的路由表中没有去往 192.168.1.110 的路由，因为 192.168.1.110 为私网地址，在公网中是不存在到私网的路由的。若配置了 Identity NAT，内部区域的主机 PC2 将不能访问外部区域。该例子只用来说明 Identity NAT 的工作原理与配置方法。

8. 配置 ACL，允许 DMZ 区域内的 ICMP 和 Telnet 报文流向内部区域

1）配置 ACL 前，从 DMZ 区域的 Server1 ping 内部区域的主机 PC2

```
Server1> ping 192.168.1.110
192.168.1.110 icmp_seq= 1 timeout
192.168.1.110 icmp_seq= 2 timeout
192.168.1.110 icmp_seq= 3 timeout
192.168.1.110 icmp_seq= 4 timeout
192.168.1.110 icmp_seq= 5 timeout
```

2）查看 ASA 日志

```
cxyASA(config)#show logging | include icmp    //只显示包含"icmp"的日志
    Oct 25 2023 16:21:06: %ASA-3-106014：Deny inbound icmp src DMZ: 172.16.2.22 dst inside :
192.168.1.110 (type 8，code 0)
    Oct 25 2023 16:21:08: %ASA-3-106014：Deny inbound icmp src DMZ: 172.16.2.22 dst inside :
192.168.1.110 (type 8，code 0)
    Oct 25 2023 16:21:10: %ASA-3-106014：Deny inbound icmp src DMZ: 172.16.2.22 dst inside :
192.168.1.110 (type 8，code 0)
    Oct 25 2023 16:21:12: %ASA-3-106014：Deny inbound icmp src DMZ: 172.16.2.22 dst inside :
192.168.1.110 (type 8，code 0)
```

Oct 25 2023 16:21:14: %ASA-3-106014：Deny inbound icmp src DMZ: 172.16.2.22 dst inside : 192.168.1.110 (type 8，code 0)

可见，从 DMZ 区域的 Server1 去往内部区域的主机 PC2 的 ICMP 流量被拒绝了。这是因为防火墙默认的安全策略是：安全级别低的区域不能访问安全级别高的区域。当然，PC2 此时是可以 ping 通 Server1 的。

3）配置 ACL，只允许 DMZ 区域内主机的 ICMP 和 Telnet 报文流向内部区域

cxyASA(config)#access-list DMZ-to-inside extended permit icmp 172.16.2.0 255.255.255.0 192.168.1.0 255.255.255.0
//允许源 IP 地址在子网 172.16.2.0/24 内的主机到目的子网 192.168.1.0/24 的 ICMP 报文
cxyASA(config)#access-list DMZ-to-inside extended permit tcp 172.16.2.0 255.255.255.0 192.168.1.0 255.255.255.0 eq telnet
//允许源 IP 地址在子网 172.16.2.0/24 内的主机到目的子网 192.168.1.0/24 的 Telnet 报文
cxyASA(config)#access-group DMZ-to-inside in interface DMZ
//在 DMZ 接口 in 方向上应用名为 DMZ-to-inside 的 ACL

4）配置 ACL 后，从 DMZ 区域的 Server1 ping 内部区域的主机 PC2

Server1>ping 192.168.1.110
84 bytes from 192.168.1.110 icmp_seq=1 ttl= 62 time =4.375 ms
84 bytes from 192.168.1.110 icmp_seq=2 ttl= 62 time =1.173 ms
84 bytes from 192.168.1.110 icmp_seq=3 ttl= 62 time =1.661 ms
84 bytes from 192.168.1.110 icmp_seq=4 ttl= 62 time =1.284 ms
84 bytes from 192.168.1.110 icmp_seq=5 ttl= 62 time =3.651 ms

5）在 Telnet 前，先在 R1 上设置允许 Telnet

cxyR1(config)#line vty 0 4
cxyR1(config-line)#password cxy   //设置登录密码
cxyR1(config-line)#login
cxyR1(config-line)#transport input telnet   // 开启 Telnet，只允许 Telnet 流量进入
cxyR1(config-line)#privilege level 15   // 设置权限等级为 15
cxyR1 (config)#enable password cxy   // 设置进入特权模式的密码

6）在 R2 上 Telnet R1

（1）Telnet 时指定源 IP 地址。

cxyR2#telnet 192.168.1.1 /source-interface e0/1   // 指定 Telnet 使用的源 IP 地址
Trying 192.168.1.1…Open

User Access Verification
Password：
cxyR1#exit

[Connection to 192.168.1.1 closed by foreign host]

（2）Telnet 时不指定源 IP 地址。

cxyR2#telnet 192.168.1.1

Trying 192.168.1.1 …

% Connection timed out; remote host not responding

可见，Telnet 时若不指定源 IP 地址，则 R2 无法 Telnet 到 R1。

（3）查看 ASA 日志。

Oct 25 2022 19:05:48: %ASA-4-106023: Deny tcp src DMZ: 172.16.100.2/42842 dst inside: 192.168.1.1/23 by access-group "DMZ-to-inside" [0x0，0x0]

Oct 25 2022 19:05:50: %ASA-4-106023: Deny tcp src DMZ: 172.16.100.2/42842 dst inside: 192.168.1.1/23 by access-group "DMZ-to-inside" [0x0，0x0]

请根据上述 ASA 日志，解释 Telnet 时不指定源 IP 地址则无法 Telnet 到 192.168.1.1 的原因。

9. 配置 ACL，禁止内部区域的 Server0 去往其他区域

1）配置 ACL 前，从内部区域的 Server0 ping 另外两个区域的主机

（1）在 Server0 上 ping DMZ 区域的 R2。

Server0>ping 172.16.2.1

84 bytes from 172.16.2.1 icmp_seq=1 ttl= 254 time =7.280 ms

84 bytes from 172.16.2.1 icmp_seq=2 ttl= 254 time =1.326 ms

84 bytes from 172.16.2.1 icmp_seq=3 ttl= 254 time =4.161 ms

84 bytes from 172.16.2.1 icmp_seq=4 ttl= 254 time =4.212 ms

84 bytes from 172.16.2.1 icmp_seq=5 ttl= 254 time =3.227 ms

（2）在 Server0 上 ping 外部区域的 R3。

Server0>ping 123.123.123.123

84 bytes from 123.123.123.123 icmp_seq=1 ttl= 254 time =6.663 ms

84 bytes from 123.123.123.123 icmp_seq=2 ttl= 254 time =2.470 ms

84 bytes from 123.123.123.123 icmp_seq=3 ttl= 254 time =3.876 ms

84 bytes from 123.123.123.123 icmp_seq=4 ttl= 254 time =4.189 ms

84 bytes from 123.123.123.123 icmp_seq=5 ttl= 254 time =2.977 ms

2）配置 ACL，实现内部区域的 Server0（192.168.1.88）只能访问本区域内的主机，不允许去往其他区域

cxyASA(config)#access-list deny-inside standard deny 192.168.1.88 255.255.255.255

cxyASA(config)#access-list deny-inside standard permit any

cxyASA(config)#access-group deny-inside in interface inside

ERROR：Only "extended" or "ethertype" acls can be attached to an interface

//注意：本步骤使用标准 ACL 是可以实现的，但 ASA 规定标准 ACL 不能在接口上应用，
//故下面用扩展 ACL 实现相同的功能
cxyASA(config)#access-list deny-inside-any extended deny ip host 192.168.1.88 any
//拒绝源 IP 地址为 192.168.1.88 的主机到任意目标主机的 IP 分组
cxyASA(config)#access-list deny-inside-any extended permit ip any any
//允许任意 IP 分组
cxyASA(config)#access-group deny-inside-any in interface inside
//在内部接口的 in 方向上应用名为 deny-inside-any 的 ACL

3）配置 ACL 后，从内部区域的 Server0 ping 另外三台主机

（1）在 Server0 上 ping DMZ 区域的 R2。

```
Server0> ping 172.16.2.1
172.16.2.1 icmp_seq= 1 timeout
172.16.2.1 icmp_seq= 2 timeout
172.16.2.1 icmp_seq= 3 timeout
172.16.2.1 icmp_seq= 4 timeout
172.16.2.1 icmp_seq= 5 timeout
```

（2）在 Server0 上 ping 外部区域的 R3。

```
Server0> ping 123.123.123.123
123.123.123.123 icmp_seq= 1 timeout
123.123.123.123 icmp_seq= 2 timeout
123.123.123.123 icmp_seq= 3 timeout
123.123.123.123 icmp_seq= 4 timeout
```

（3）在 Server0 上 ping 内部区域的 R1。

```
Server0> ping 192.168.100.1
84 bytes from 192 .168.100.1 icmp_seq=1 ttl= 255 time=0.519 ms
84 bytes from 192 .168.100.1 icmp_seq=2 ttl= 255 time=0.378 ms
84 bytes from 192 .168.100.1 icmp_seq=3 ttl= 255 time=0.428 ms
84 bytes from 192 .168.100.1 icmp_seq=4 ttl= 255 time=0.612 ms
84 bytes from 192 .168.100.1 icmp_seq=5 ttl= 255 time=0.408 ms
```

第 16 章 策略路由

16.1 组网需求

网络拓扑如图 16-1 所示，其中，SW1 为 Cisco IOSvL2 2020 系列交换机，现要求实现以下功能。

（1）配置 OSPF 使得全网在 IP 层互通。

（2）VLAN10 的 IP 地址范围为 176.16.60.0～176.16.100.255。请用尽可能少的 ACL 规则，实现策略路由，出口为 R1。

（3）VLAN20 的 IP 地址范围为 180.10.2.0/24。要求：

①180.10.2.64/27 范围内且为奇数的 IP 地址转发到 R1，176.16.1.64/27 范围内且为偶数的 IP 地址转发到 R2。

②为提高可靠性，在 180.10.2.64/27 范围内奇、偶 IP 地址的互联网访问互为备份。

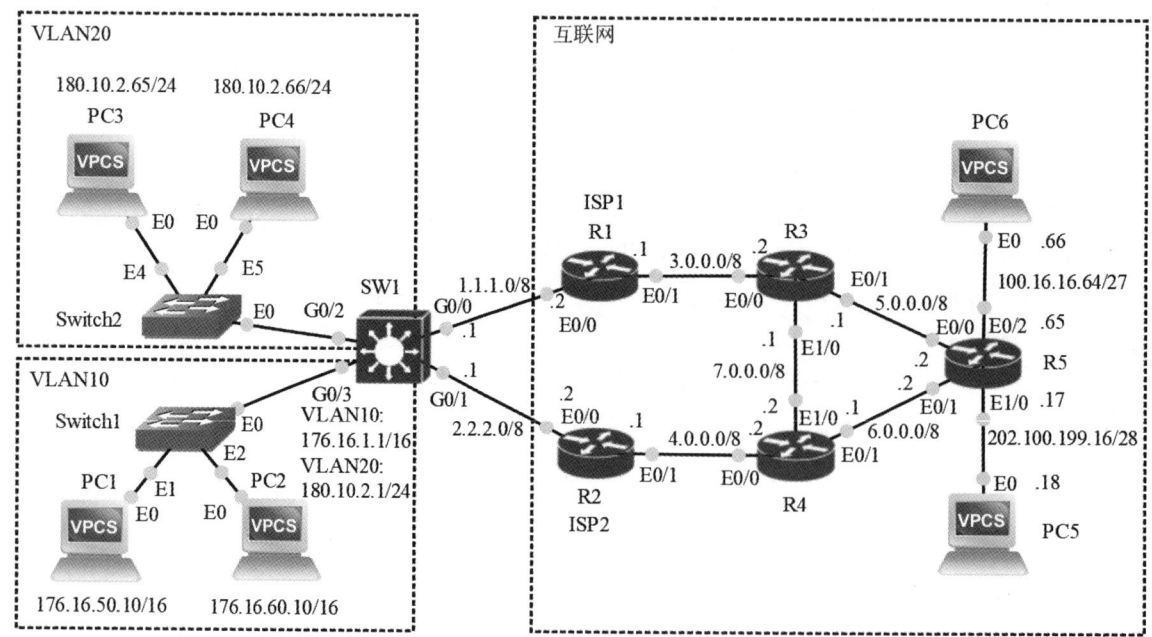

图 16-1 网络拓扑

16.2 仿真实验

16.2.1 仿真环境

仿真环境所需要的硬件、软件建议至少满足以下要求。

- 物理机的处理器主频为 1.6GHz；
- 内存 RAM 大小为 32.0GB；
- 物理机的操作系统为 Windows 10；
- 虚拟机为 VMware Workstation 16 Pro；
- 网络模拟器选用 GNS3 2.2.31；
- 核心交换机为 Cisco IOSvL2 2020；
- 二层交换机为 Ethernet Switch；
- 路由器为 Cisco IOL Router15.7；
- PC1~PC6 为 Virtual PC Simulator, Version 0.8.2。

16.2.2 实验设计

根据网络拓扑（图 16-1）完成网络设备的选型：1 台三层交换机选用 Cisco IOSvL2 2020 系列交换机，2 台二层交换机选用 Ethernet Switch，5 台路由器选用 Cisco IOL Router15.7 系列路由器，6 台 PC 选用 VPCS。根据前面的组网需求，本实验设计如下。

1. VLAN 设计

在三层交换机 SW1 上创建两个 VLAN：VLAN10 和 VLAN20，并将接口 G0/3 划分到 VLAN10 中，将接口 G0/2 划分到 VLAN20 中。

2. IP 地址规划

三层交换机和路由器各接口的 IP 地址/掩码如表 16-1 所示。

表 16-1　各设备接口的 IP 地址/掩码

设备	接口	IP 地址/掩码
R1	E0/0	1.1.1.2/8
R1	E0/1	3.0.0.1/8
R2	E0/0	2.2.2.2/8
R2	E0/1	4.0.0.1/8
R3	E0/0	3.0.0.2/8
R3	E0/1	5.0.0.1/8
R3	E1/0	7.0.0.1/8
R4	E0/0	4.0.0.2/8
R4	E0/1	6.0.0.1/8
R4	E1/0	7.0.0.2/8
R5	E0/0	5.0.0.2/8
R5	E0/1	6.0.0.2/8
R5	E0/2	100.16.16.65/27

续表

设备	接口	IP 地址/掩码
R5	E1/0	202.100.199.17/28
SW1	G0/0	1.1.1.1/8
SW1	G0/1	2.2.2.1/8
SW1	VLAN10	176.16.1.1/16
SW1	VLAN20	180.10.2.1/24

3. 策略路由设计

通过策略路由控制访问互联网的出口，实现 VLAN10 中 IP 地址范围为 176.16.60.0～176.16.100.255 的主机访问互联网时出口为 R1。在配置策略路由时，要求使用尽可能少的 ACL 规则。

通过策略路由实现基于源 IP 地址的路由选择，要求源 IP 地址在 180.10.2.64/27 范围内且为奇数的分组转发到 R1，而源 IP 地址在 180.10.2.64/27 范围内且为偶数的分组转发到 R2。

通过配置策略路由，使得 180.10.2.64/27 范围内奇、偶 IP 地址的互联网访问互为备份，即可提高内网访问互联网的可靠性。

16.2.3 操作步骤

下面给出本次实验的具体操作，一共包括 9 个步骤。

1. 按照图 16-1 所示的网络拓扑，在 GNS3 中新建一个空白的工程

网络设备选型如下：交换机 SW1 选择 Cisco IOSvL2 2020，R1～R5 选择 Cisco IOL Router 15.7，Switch1、Switch2 选择 GNS3 自带的二层交换机。

2. 配置三层交换机 SW1

1）创建 VLAN 并将接口划分到相应的 VLAN 中

```
SW1(config)#hostname cxySW1                         //可将 cxy 换成你的名字拼音首字母
cxySW1(config)#vlan10                               //创建 VLAN10 和 VLAN20
cxySW1(config-vlan)#vlan20
cxySW1(config-vlan)#exit
cxySW1(config)#interface g0/2
cxySW1(config-if)#switchport access vlan20          //将接口 g0/2 加入 VLAN20 中
cxySW1(config-if)# interface g0/3
cxySW1(config-if)#switchport access vlan10          //将接口 g0/3 加入 VLAN10 中
```

2）查看 VLAN 信息

```
cxySW1#show vlan
VLAN   Name        Status     Ports
1      default     active     Gi0/1，Gi1/0，Gi1/1，Gi1/2
                              Gi1/3，Gi2/0，Gi2/1，Gi2/2
                              Gi2/3，Gi3/0，Gi3/1，Gi3/2
```

			Gi3/3
10	VLAN0010	active	Gi0/3
20	VLAN0020	active	Gi0/2
1002	fddi-default	act/unsup	
1003	token-ring-default	act/unsup	
1004	fddinet-default	act/unsup	
1005	trnet-default	act/unsup	

VLAN	Type	SAID	MTU	Parent	RingNo	BridgeNo	Stp	BrdgMode	Trans1	Trans2
1	enet	100001	1500	-	-	-	-	-	0	0
10	enet	100010	1500	-	-	-	-	-	0	0
20	enet	100020	1500	-	-	-	-	-	0	0
1002	fddi	101002	1500	-	-	-	-	-	0	0
1003	tr	101003	1500	-	-	-	-	-	0	0
1004	fdnet	101004	1500	-	-	-	ieee	-	0	0
1005	trnet	101005	1500	-	-	-	ibm	-	0	0

Remote SPAN VLANs

Primary	Secondary	Type	Ports

3）配置 SW1 接口 IP 地址

（1）配置 VLAN 接口 IP 地址。

cxySW1(config)#interface vlan10
cxySW1(config-if)#ip address 176.16.1.1 255.255.0.0
cxySW1(config-if)#no shutdown
*Nov 6 01:10:16.507: %LINK-3-UPDOWN：Interface Vlan10，changed state to up
*Nov 6 01:10:17.508: %LINEPROTO-5-UPDOWN：Line protocol on Interface Vlan10, changed state to up
cxySW1(config-if)#interface vlan20
cxySW1(config-if)#ip address 180.10.2.1 255.255.255.0
cxySW1(config-if)#no shutdown
*Nov 6 01:13:51.315: %LINK-3-UPDOWN：Interface Vlan20，changed state to up
*Nov 6 01:13:52.315: %LINEPROTO-5-UPDOWN: Line protocol on Interface Vlan20, changed state to up

（2）启用交换机三层接口功能并配置 IP 地址。

cxySW1(config-if)#interface g0/0

```
cxySW1(config-if)#no switchport
cxySW1(config-if)#ip address 1.1.1.1 255.0.0.0
cxySW1(config-if)#no shutdown
cxySW1(config-if)#interface g0/1
cxySW1(config-if)#no switchport
cxySW1(config-if)#ip address 2.2.2.1 255.0.0.0
cxySW1(config-if)#no shutdown
```

（3）查看路由表。

```
cxySW1#show ip route
Codes：L-local，C-connected，S-static，R-RIP，M-mobile，B-BGP
    D-EIGRP，EX-EIGRP external，O-OSPF，I A-OSPF inter area
    N1-OSPF NSSA external type 1，N2-OSPF NSSA external type 2
    E1-OSPF external type1，E2-OSPF external type 2
    i-IS-IS，su-IS-IS summary，L1-IS-IS level-1，L2-IS-IS level-2
    ia-IS-IS inter area，*- candidate default，U-per-user static route
    o-ODR，P-periodic downloaded static route，H-NHRP，l-LISP
    a-application route
    +-replicated route，%-next hop override，p-overrides from PfR
Gateway of last resort is not set
         1.0.0.0/8 is variably subnetted，2 subnets，2 masks
C        1.0.0.0/8 is directly connected，GigabitEthernet 0/0
L        1.1.1.1/32 is directly connected，GigabitEthernet 0/0
         2.0.0.0/8 is variably subnetted，2 subnets，2 masks
C        2.0.0.0/8 is directly connected，GigabitEthernet 0/1
L        2.2.2.1/32 is directly connected，GigabitEthernet 0/1
         176.16.0.0/16 is variably subnetted，2 subnets，2 masks
C        176.16.0.0/16 is directly connected，Vlan10
L        176.16.1.1/32 is directly connected，Vlan10
         180.10.0.0/16 is variably subnetted，2 subnets，2 masks
C        180.10.2.0/24 is directly connected，Vlan20
L        180.10.2.1/32 is directly connected，Vlan20
```

3. 配置二层交换机

1）配置二层交换机 Switch1

在 GNS3 拓扑图中，右击交换机 Switch1 的图标，在弹出的菜单中选择"Configure"命令，弹出如图 16-2 所示的"Node properties"窗口；在"Node properties"窗口中，将 Switch1 的 E0、E1、E2 接口加入 VLAN10 中，配置结果如图 16-2 中的"Ports"列表框所示。

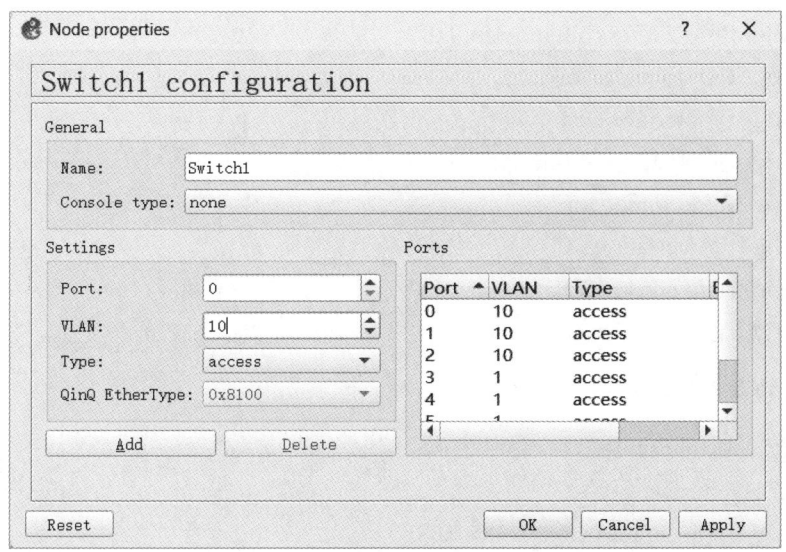

图 16-2 "Node properties" 窗口

2）配置二层交换机 Switch2

按照前面配置二层交换机 Switch1 的方法，将 Switch2 的 E0、E4、E5 接口加入 VLAN20 中。

4. 配置路由器接口及 PC 的 IP 地址

1）配置路由器 R1 接口的 IP 地址

R1(config)#hostname cxyR1

cxyR1(config)#interface e0/0

cxyR1(config-if)#ip address 1.1.1.2 255.0.0.0

cxyR1(config-if)#no shutdown

cxyR1(config-if)#interface e0/1

cxyR1(config-if)#ip address 3.0.0.1 255.0.0.0

cxyR1(config-if)#no shutdown

2）配置路由器 R2 接口的 IP 地址

R2(config)#hostname cxyR2

cxyR2(config-if)#interface e0/0

cxyR2(config-if)#ip address 2.2.2.2 255.0.0.0

cxyR2(config-if)#no shutdown

cxyR2(config-if)#interface e0/1

cxyR2(config-if)#ip address 4.0.0.1 255.0.0.0

cxyR2(config-if)#no shutdown

3）配置路由器 R3 接口的 IP 地址

R3(config)#hostname cxyR3

cxyR3(config)#interface e0/0

cxyR3(config-if)#ip address 3.0.0.2 255.0.0.0
cxyR3(config-if)#no shutdown
cxyR3(config-if)#interface e0/1
cxyR3(config-if)#ip address 5.0.0.1 255.0.0.0
cxyR3(config-if)#no shutdown
cxyR3(config-if)#interface e1/0
cxyR3(config-if)#ip address 7.0.0.1 255.0.0.0
cxyR3(config-if)#no shutdown

4）配置路由器 R4 接口的 IP 地址

R4(config)#hostname cxyR4
cxyR4(config)#interface e0/0
cxyR4(config-if)#ip address 4.0.0.2 255.0.0.0
cxyR4(config-if)#no shutdown
cxyR4(config-if)#interface e0/1
cxyR4(config-if)#ip address 6.0.0.1 255.0.0.0
cxyR4(config-if)#no shutdown
cxyR4(config-if)#interface e1/0
cxyR4(config-if)#ip address 7.0.0.2 255.0.0.0
cxyR4(config-if)#no shutdown

5）配置路由器 R5 接口的 IP 地址

R5(config)#hostname cxyR5
cxyR5(config-if)#interface e0/0
cxyR5(config-if)#ip address 5.0.0.2 255.0.0.0
cxyR5(config-if)#no shutdown
cxyR5(config)#interface e0/1
cxyR5(config-if)#ip address 6.0.0.2 255.0.0.0
cxyR5(config-if)#no shutdown
cxyR5(config)#interface e0/2
cxyR5(config-if)#ip address 100.16.16.65 255.255.255.224
cxyR5(config-if)#no shutdown
cxyR5(config-if)#interface e1/0
cxyR5(config-if)#ip address 202.100.199.17 255.255.255.240
cxyR5(config-if)#no shutdown

6）配置 PC1～PC6 的 IP 地址及网关

PC1> ip 176.16.50.10/16 176.16.1.1
PC2> ip 176.16.60.10/16 176.16.1.1
PC3> ip 180.10.2.65/24 180.10.2.1

PC4> ip 180.10.2.66/24 180.10.2.1
PC5> ip 202.100.199.18/28 202.100.199.17
PC6> ip 100.16.16.66/27 100.16.16.65

5. 配置静态路由和 OSPF

1）配置三层交换机的静态路由和 OSPF

cxySW1(config)#ip route 100.16.16.64 255.255.255.224 1.1.1.2
cxySW1(config)#ip route 202.100.199.16 255.255.255.240 2.2.2.2
cxySW1(config)#router ospf 10
cxySW1(config-router)#network 1.0.0.0 0.255.255.255 area 0
cxySW1(config-router)#network 2.0.0.0 0.255.255.255 area 0

2）配置路由器 R1 的静态路由和 OSPF

cxyR1(config)#ip route 0.0.0.0 0.0.0.0 1.1.1.1
cxyR1(config)#router ospf 10
cxyR1(config-router)#network 1.0.0.0 0.255.255.255 area 0
cxyR1(config-router)#network 3.0.0.0 0.255.255.255 area 0

3）配置路由器 R2 的静态路由和 OSPF

cxyR2(config)#ip route 0.0.0.0 0.0.0.0 2.2.2.1
cxyR2(config)#router ospf 10
cxyR2(config-router)#network 2.0.0.0 0.255.255.255 area 0
cxyR2(config-router)#network 4.0.0.0 0.255.255.255 area 0

4）配置路由器 R3 的静态路由和 OSPF

cxyR3(config)#ip route 0.0.0.0 0.0.0.0 3.0.0.1
cxyR3(config)#router ospf 10
cxyR3(config-router)#network 3.0.0.0 0.255.255.255 area 0
cxyR3(config-router)#network 5.0.0.0 0.255.255.255 area 0
cxyR3(config-router)#network 7.0.0.0 0.255.255.255 area 0

5）配置路由器 R4 的静态路由和 OSPF

cxyR4(config)#ip route 0.0.0.0 0.0.0.0 4.0.0.1
cxyR4(config)#router ospf 10
cxyR4(config-router)#network 4.0.0.0 0.255.255.255 area 0
cxyR4(config-router)#network 6.0.0.0 0.255.255.255 area 0
cxyR4(config-router)#network 7.0.0.0 0.255.255.255 area 0

6）配置路由器 R5 的静态路由和 OSPF

cxyR5(config)#ip route 0.0.0.0 0.0.0.0 5.0.0.1

```
cxyR5(config)#ip route 0.0.0.0 0.0.0.0 6.0.0.1
cxyR5(config)#router ospf 10
cxyR5(config-router)#network 5.0.0.0 0.255.255.255 area 0
cxyR5(config-router)#network 6.0.0.0 0.255.255.255 area 0
cxyR5(config-router)#network 202.100.199.16 0.0.0.15 area 0
cxyR5(config-router)# network 100.16.16.64 0.0.0.31 area 0
```

6. 查看路由表及路由跟踪

1）查看路由器 R5 的路由表

```
cxyR5#show ip route
Codes: L-local, C-connected, S-static, R-RIP, M-mobile, B-BGP
       D-EIGRP, EX-EIGRP external, O-OSPF, I A-OSPF inter area
       N1-OSPF NSSA external type 1, N2-OSPF NSSA external type 2
       E1-OSPF external type1, E2-OSPF external type 2
       i-IS-IS, su-IS-IS summary, L1-IS-IS level-1, L2-IS-IS level-2
       ia-IS-IS inter area, *- candidate default, U-per-user static route
       o-ODR, P-periodic downloaded static route, H-NHRP, l-LISP
       a-application route
       +-replicated route, %-next hop override, p-overrides from PfR
Gateway of last resort is 6.0.0.1 to network 0.0.0.0
s*   0.0.0.0/0 [1/0] via 6.0.0.1
                [1/0] via 5.0.0.1
O    1.0.0.0/8 [110/31] via 6.0.0.1, 06:20:25, Ethernet 0/1
O    2.0.0.0/8 [110/30] via 6.0.0.1, 12:01:27, Ethernet 0/1
O    3.0.0.0/8 [110/20] via 5.0.0.1, 12:01:27, Ethernet 0/0
O    4.0.0.0/8 [110/20] via 6.0.0.1, 12:01:27, Ethernet 0/1
     5.0.0.0/8 is variably subnetted, 2 subnets, 2 masks
C    5.0.0.0/8 is directly connected, Ethernet 0/0
L    5.0.0.2/32 is directly connected, Ethernet 0/0
     6.0.0.0/8 is variably subnetted, 2 subnets, 2 masks
C    6.0.0.0/8 is directly connected, Ethernet 0/1
L    6.0.0.2/32 is directly connected, Ethernet 0/1
O    7.0.0.0/8 [110/20] via 6.0.0.1, 12:01:27, Ethernet 0/1
               [110/20] via 5.0.0.1, 12:01:27, Ethernet 0/0
     100.0.0.0/8 is variably subnetted, 2 subnets, 2 masks
C    100.16.16.64/27 is directly connected, Ethernet 0/2
L    100.16.16.65/32 is directly connected, Ethernet 0/2
     202.100.199.0/24 is variably subnetted, 2 subnets, 2 masks
C    202.100.199.16/28 is directly connected, Ethernet 1/0
```

L   202.100.199.17/32 is directly connected，Ethernet 1/0

值得注意的是，R5 的路由表中存在两条到达目的网络 7.0.0.0/8 的等值路由。

2）查看三层交换机 SW1 的路由表

cxySW1#show ip route
Codes：L-local，C-connected，S-static，R-RIP，M-mobile，B-BGP
　　D-EIGRP，EX-EIGRP external，O-OSPF，I A-OSPF inter area
　　N1-OSPF NSSA external type 1，N2-OSPF NSSA external type 2
　　E1-OSPF external type1，E2-OSPF external type 2
　　i-IS-IS，su-IS-IS summary，L1-IS-IS level-1，L2-IS-IS level-2
　　ia-IS-IS inter area，*- candidate default，U-per-user static route
　　o-ODR，P-periodic downloaded static route，H-NHRP，l-LISP
　　a-application route
　　+-replicated route，%-next hop override，p-overrides from PfR
Gateway of last resort is not set

　　　1.0.0.0/8 is variably subnetted，2 subnets，2 masks
C　　1.0.0.0/8 is directly connected，GigabitEthernet 0/0
L　　1.1.1.1/32 is directly connected，GigabitEthernet 0/0
　　　2.0.0.0/8 is variably subnetted，2 subnets，2 masks
C　　2.0.0.0/8 is directly connected，GigabitEthernet 0/1
L　　2.2.2.1/32 is directly connected，GigabitEthernet 0/1
O　　3.0.0.0/8 [110/11] via 1.1.1.2，00:22:24，GigabitEthernet 0/0
O　　4.0.0.0/8 [110/11] via 2.2.2.2，12:35:53，GigabitEthernet 0/1
O　　5.0.0.0/8 [110/21] via 1.1.1.2，00:22:24，GigabitEthernet 0/0
O　　6.0.0.0/8 [110/21] via 2.2.2.2，12:32:40，GigabitEthernet 0/1
O   **7.0.0.0/8 [110/21] via 2.2.2.2，12:32:40，GigabitEthernet 0/1**
　　　　　[110/21] via 1.1.1.2，00:22:24，GigabitEthernet 0/0
　　　100.0.0.0/27 is subnetted，1 subnets
S　　100.16.16.64 [1/0] via 1.1.1.2
　　　176.16.0.0/16 is variably subnetted，2 subnets，2 masks
C　　176.16.0.0/16 is directly connected，Vlan 10
L　　176.16.1.1/32 is directly connected，Vlan 10
　　　180.10.0.0/16 is variably subnetted，2 subnets，2 masks
C　　180.10.2.0/24 is directly connected，Vlan 20
L　　180.10.2.1/32 is directly connected，Vlan 20
　　　202.100.199.0/28 is subnetted，1 subnets
S　　202.100.199.16 [1/0] via 2.2.2.2

值得注意的是，三层交换机 SW1 的路由表中存在两条到达目的网络 7.0.0.0/8 的等值路由。等值路由在其他一些路由中也存在，可通过下面的路由跟踪加深理解。

3）在 PC 上进行路由跟踪

（1）从 PC6 跟踪到达 7.0.0.1 的路由。

```
PC6> trace 7.0.0.1
trace to 7.0.0.1, 8 hops max, press Ctrl+C to stop
1   100.16.16.65    0.169ms         0.123ms         0.052ms
2   6.0.0.1         0.172ms         0.156ms         0.107ms
3   *7.0.0.1        0.213ms (ICMP type: 3, code: 3, Destination port unreachable)
```

（2）从 PC6 跟踪到达 7.0.0.2 的路由。

```
PC6> trace 7.0.0.2
trace to 7.0.0.2, 8 hops max, press Ctrl+C to stop
1   100.16.16.65    0.133ms         0.111ms         0.059ms
2   5.0.0.1         0.165ms         0.108ms         0.141ms
3   *7.0.0.2        0.378ms (ICMP type: 3, code: 3, Destination port unreachable)
```

可见，从 PC6 到达网络 7.0.0.0/8 的 ICMP 报文从 R5 开始分别走不同的路径了。

（3）从 PC1 跟踪到达 202.100.199.18 的路由。

```
PC1> trace 202.100.199.18
trace to 202.100.199.18, 8 hops max, press Ctrl+C to stop
1   176.16.1.1      2.658ms         2.160ms         1.699ms
2   2.2.2.2         2.971ms         4.741ms         5.440ms
3   4.0.0.2         3.449ms         6.466ms         4.868ms
4   6.0.0.2         5.179ms         4.531ms         4.718ms
5   *202.100.199.18 7.382ms (ICMP type: 3, code: 3, Destination port unreachable)
```

（4）从 PC2 跟踪到达 202.100.199.18 的路由。

```
PC2> trace 202.100.199.18
trace to 202.100.199.18, 8 hops max, press Ctrl+C to stop
1   176.16.1.1      1.741ms         2.126ms         2.084ms
2   2.2.2.2         4.369ms         4.292ms         4.201ms
3   4.0.0.2         3.768ms         2.869ms         4.813ms
4   6.0.0.2         3.796ms         2.194ms         2.386ms
5   *202.100.199.18 1.408ms (ICMP type: 3, code: 3, Destination port unreachable)
```

由此可见：PC1、PC2 访问 202.100.199.18 时，走的是相同的路径。此时的路径是由路由表决定的。

（5）从 PC2 跟踪到达 180.10.2.65 的路由。

```
PC2> trace 180.10.2.65
trace to 180.10.2.65, 8 hops max, press Ctrl+C to stop
```

1	176.16.1.1	2.193ms 2.375ms	1.291ms
2	*180.10.2.65	6.543ms (ICMP type：3，code：3，Destination port unreachable)	

不难看出：从 PC2 访问 PC3 是通过 SW1 的直连路由实现的。

7. 通过策略路由控制访问互联网的出口

通过策略路由控制访问互联网的出口，实现 VLAN10 内 IP 地址范围为 176.16.60.0～176.16.100.255 的主机访问互联网时出口为 R1。在配置策略路由时，要求使用尽可能少的 ACL 规则。

1）定义 ACL，允许源 IP 地址范围为 176.16.60.0～176.16.100.255 的主机访问互联网

cxySW1(config)#access-list 1 permit 176.16.60.0 0.0.3.255
cxySW1(config)#access-list 1 permit 176.16.64.0 0.0.31.255
cxySW1(config)#access-list 1 permit 176.16.96.0 0.0.3.255
cxySW1(config)#access-list 1 permit 176.16.100.0 0.0.0.255

2）查看 ACL

cxySW1#show access-list 1
Standard IP access list 1
10 permit 176.16.60.0，wildcard bits 0.0.3.255
20 permit 176.16.64.0，wildcard bits 0.0.31.255
30 permit 176.16.96.0，wildcard bits 0.0.3.255
40 permit 176.16.100.0，wildcard bits 0.0.0.255

3）配置名为"176.16.60.0toR1"的策略路由

cxySW1(config)#route-map 176.16.60.0toR1 permit 10   //策略序号为 10
cxySW1(config-route-map)#match ip address 1   //引用 ACL 1
cxySW1(config-route-map)#set ip next-hop 1.1.1.2   //设置下一跳为 R1

4）查看策略路由

cxySW1#show route-map
   route-map 176.16.60.0toR1，permit，sequence10
Match clauses：
   ip address (access-lists)：1
Set clauses：
   ip next-hop 1.1.1.2
policy routing matches：44 packets，4624 bytes

5）在 VLAN10 接口上应用策略路由

cxySW1(config)#interface vlan10
cxySW1(config-if)#ip policy route-map 176.16.60.0toR1   //应用策略路由 176.16.60.0toR1

6）调试策略路由

（1）在 PC2 上跟踪到达 202.100.199.18 的路由。

```
PC2> trace 202.100.199.18
trace to 202.100.199.18，8 hops max，press Ctrl+C to stop
1    176.16.1.1       2.948ms        2.942ms        1.686ms
2    1.1.1.2          1.827ms        2.128ms        2.718ms
3    3.0.0.2          1.959ms        3.845ms        2.263ms
4    5.0.0.2          2.323ms        2.032ms        1.625ms
5    *202.100.199.18  4.298ms (ICMP type：3，code：3，Destination port unreachable)
```

可见，应用策略路由前后，访问互联网时出口发生了变化，达到了选择出口的目的。

（2）在 PC2 上跟踪到达 180.10.2.65 的路由。

```
PC2> trace 180.10.2.65
trace to 180.10.2.65，8 hops max，press Ctrl+Cto stop
1    176.16.1.1       3.213ms        2.552ms        1.303ms
2    1.1.1.2          2.352ms        2.722ms        1.873ms
3    1.1.1.1          2.832ms        2.188ms        3.836ms
4    *180.10.2.65     6.518ms (ICMP type：3，code：3，Destination port unreachable)
```

请比较应用策略路由前后，PC2 访问 PC3（180.10.2.65）时路径的变化，并解释其原因。若要求通过 SW1 的直连路由转发从 VLAN10 到 VLAN20 的流量，应该如何实现？能否通过 set default next-hop 命令实现？

7）在"176.16.60.0toR1"策略路由中增加一条序号为 1 的策略

（1）定义目的为 VLAN20 的 ACL，用于策略路由。

```
cxySW1(config)#access-list 101 permit ip any 180.10.2.0 0.0.0.255
```

（2）增加一条序号为 1 的策略。

```
cxySW1(config)#route-map 176.16.60.0toR1 deny 1              //策略序号为 1
cxySW1(config-route-map)#match ip address 101                //引用 ACL 101
```

（3）查看策略路由。

```
cxySW1#show route-map
    route-map 176.16.60.0toR1，deny，sequence 1
Match clauses：
    ip address (access-lists)：101
Set clauses：
Policy routing matches：0 packets，0 bytes
    route-map 176.16.60.0toR1，permit，sequence 10
Match clauses：
```

```
        ip address (access-lists): 1
Set clauses:
        ip next-hop 1.1.1.2
Policy routing matches: 44 packets, 4624 bytes
```

需要注意的是：序号越小的策略优先级越高。

（4）在 PC2 上跟踪到达 180.10.2.65 的路由。

```
PC2> trace 180.10.2.65
trace to 180.10.2.65, 8 hops max, press Ctrl+C to stop
1   176.16.1.1         1.933ms                2.421ms                  1.334ms
2   *180.10.2.65       4.106ms (ICMP type: 3, code: 3, Destination port unreachable)
```

可见，此时从 VLAN10 到 VLAN20 的流量是通过 SW1 的直连路由转发的。

8. 通过策略路由实现基于源 IP 地址的路由选择

要求源 IP 地址在 180.10.2.64/27 范围内且为奇数的分组转发到 R1，而源 IP 地址在 180.10.2.64/27 范围内且为偶数的分组转发到 R2。

1）定义 ACL

```
cxySW1(config)#access-list 2 permit 180.10.2.64 0.0.0.30
//定义源 IP 地址在 180.10.2.64～180.10.2.95 范围内的偶数 IP 地址的标准编号 ACL（名为 2）
cxySW1(config)#access-list 3 permit 180.10.2.65 0.0.0.30
//定义源 IP 地址在 180.10.2.64～180.10.2.95 范围内的奇数 IP 地址的标准编号 ACL（名为 3）
cxySW1(config)#access-list 100 permit ip any 176.16.0.0 0.0.255.255
//定义目的为 VLAN10 的扩展编号 ACL（名为 100）
```

2）配置策略路由

创建名为 "180toR1R2" 的策略路由，此路由包含 3 条策略。

（1）配置序号为 1 的策略。

```
cxySW1(config)#route-map 180toR1R2 deny 1
cxySW1(config-route-map)#match ip address 100
```

此策略的作用是：目的 IP 地址为 176.16.0.0/16 的分组通过普通路由表，实现 VLAN20 到 VLAN10 的数据流经 SW1 直接通信，而不是先到外网再返回来。当内网使用动态 NAT 时，外网将无内网的路由，必须配置此策略，否则 VLAN10 与 VLAN20 间无法通信。

（2）配置序号为 10 的策略。

```
cxySW1(config)#route-map 180toR1R2 permit 10
cxySW1(config-route-map)#match ip address 2
cxySW1(config-route-map)#set ip next-hop 2.2.2.2
```

（3）配置序号为 20 的策略。

```
cxySW1(config-route-map)#route-map 180toR1R2 permit 20
```

```
cxySW1(config-route-map)#match ip address 3
cxySW1(config-route-map)#set ip next-hop 1.1.1.2
```

3）在接口上应用策略路由

```
//在 VLAN20 接口上应用名为 "180toR1R2" 的策略路由
cxySW1(config)#interface vlan20
cxySW1(config-if)#ip policy route-map 180toR1R2
```

4）调试策略路由

（1）在 PC3 上跟踪到达 176.16.50.10 的路由。

```
PC3> trace 176.16.50.10
trace to 176.16.50.10, 8 hops max, press Ctrl+C to stop
 1   180.10.2.1        2.740ms              2.229ms              0.865ms
 2   *176.16.50.10     5.601ms (ICMP type: 3, code: 3, Destination port unreachable)
```

（2）在 PC3 上跟踪到达 202.100.199.18 的路由。

```
PC3> trace 202.100.199.18
trace to 202.100.199.18, 8 hops max, press Ctrl+C to stop
 1   180.10.2.1        2.197ms              2.296ms              2.764ms
 2   1.1.1.2           2.138ms              2.581ms              2.609ms
 3   3.0.0.2           2.216ms              1.918ms              2.351ms
 4   5.0.0.2           2.049ms              1.804ms              2.753ms
 5   *202.100.199.18   3.772ms (ICMP type: 3, code: 3, Destination port unreachable)
```

（3）在 PC4 上跟踪到达 202.100.199.18 的路由。

```
PC4> trace 202.100.199.18
trace to 202.100.199.18, 8 hops max, press Ctrl+C to stop
 1   180.10.2.1        3.958ms              1.737ms              1.146ms
 2   2.2.2.2           2.976ms              2.345ms              2.459ms
 3   4.0.0.2           2.552ms              2.118ms              3.680ms
 4   6.0.0.2           6.021ms              2.893ms              4.161ms
 5   *202.100.199.18   3.642ms (ICMP type: 3, code: 3, Destination port unreachable)
```

可见，应用策略路由 "180toR1R2" 后，源 IP 地址奇偶性不同的两台主机 PC3 和 PC4 在访问外网主机时，分别选择了不同的下一跳路由器。可以通过下面的对比实验加深理解。

（4）取消策略路由的对比实验。

```
cxySW1(config-route-map)#interface vlan20
cxySW1(config-if)#no ip policy route-map 180toR1R2
PC3> trace 202.100.199.18
trace to 202.100.199.18, 8 hops max, press Ctrl+C to stop
```

1	180.10.2.1	1.924ms	2.418ms	1.916ms
2	2.2.2.2	2.559ms	2.589ms	2.226ms
3	4.0.0.2	2.359ms	2.003ms	2.062ms
4	6.0.0.2	3.955ms	2.728ms	2.118ms
5	*202.100.199.18	4.812ms (ICMP type：3，code：3，Destination port unreachable)		

PC4> trace 202.100.199.18

trace to 202.100.199.18，8 hops max，press Ctrl+C to stop

1	180.10.2.1	1.991ms	1.727ms	0.854ms
2	2.2.2.2	1.814ms	2.366ms	1.609mS
3	4.0.0.2	2.142ms	2.362ms	1.958ms
4	6.0.0.2	1.875ms	2.403ms	4.546ms
5	*202.100.199.18	3.823ms (ICMP type：3，code：3，Destination port unreachable)		

可见，不应用策略路由时，SW1 是基于路由表转发分组的。对比实验完成后，恢复在接口上应用策略路由：

```
cxySW1(config-route-map)#interface vlan20
cxySW1(config-if)#ip policy route-map 180toR1R2
```

9. 通过策略路由实现可靠性的提高

通过配置策略路由，使得 180.10.2.64/27 范围内奇、偶 IP 地址的互联网访问互为备份，即可提高内网访问互联网的可靠性。

1）关闭 SW1 的接口 G0/0（模拟 SW1 的接口 G0/0 出现故障）

```
cxySW1(config)#interface g0/0
cxySW1(config-if)#shutdown
```

2）PC3 将不能访问互联网

PC3> trace 202.100.199.18

trace to 202.100.199.18，8 hops max，press Ctrl+C to stop

1	180.10.2.1	2.269ms	1.198ms	1.814ms
2	*	*	*	
3	*	*	*	
4	*	*	*	

3）配置策略路由，使互联网访问互为备份

```
cxySW1(config)#route-map 180toR1R2 permit 10
cxySW1(config-route-map)#match ip address 2
cxySW1(config-route-map)#set ip next-hop 2.2.2.2 1.1.1.2   //R2 为备份
cxySW1(config-route-map)#route-map 180toR1R2 permit 20
cxySW1(config-route-map)#match ip address 3
cxySW1(config-route-map)#set ip next-hop 1.1.1.2 2.2.2.2   //R1 为备份
```

4）通过策略路由备份，PC3 可以访问互联网

```
PC3> trace 202.100.199.18
trace to 202.100.199.18, 8 hops max, press Ctrl+C to stop
 1   180.10.2.1      2.338ms         2.099ms         1.587ms
 2   2.2.2.2         1.246ms         2.562ms         1.326ms
 3   4.0.0.2         1.675ms         1.248ms         1.622ms
 4   6.0.0.2         1.587ms         1.496ms         3.357ms
 5*  202.100.199.18  3.241ms (ICMP type：3，code：3，Destination port unreachable)
```

第 17 章 基于 ASA 防火墙的 IPsec VPN 配置

17.1 组网需求

网络拓扑如图 17-1 所示,其中 ASA1 和 ASA2 均为 Cisco ASA 系列防火墙,现要求在防火墙 ASA1 和 ASA2 之间配置站点到站点的 IPsec VPN,并对 192.168.1.0/24 和 192.168.241.0/24 网络之间的流量加密。IPsec VPN 参数值如下:加密算法选择 AES,Hash 算法选择 SHA,Diffie-Hellman 组 ID 为 14;设备认证使用预共享密钥,密钥定义为 cxyuserpass;其他参数选择默认值。

在此基础上,要求通过 Cloud 实现 GNS3 中的主机访问其他网络:实现 GNS3 中的主机 PC2 访问互联网;分公司的 TEST PC 通过浏览器访问总公司的 Web 服务器 Server。图 17-1 中把路由器 R3 模拟成 Web 服务器 Server,然后通过 TEST PC 上的浏览器访问该服务器。此时,可通过 Web 方法配置路由器 R3。

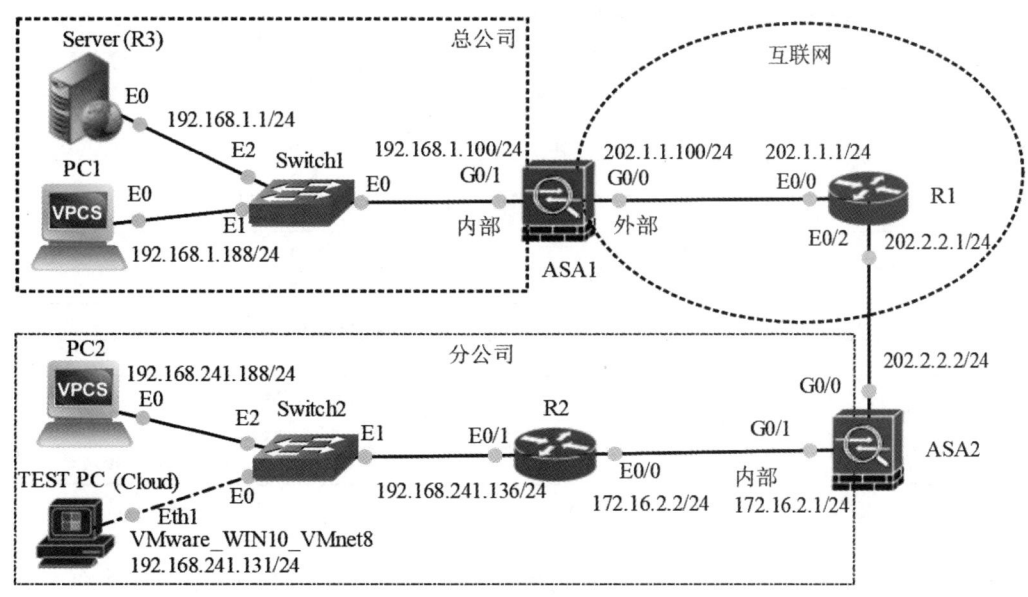

图 17-1 网络拓扑

17.2 仿真实验

17.2.1 仿真环境

仿真环境所需要的硬件、软件建议至少满足以下要求。

- 物理机的处理器主频为 1.6GHz；
- 内存 RAM 大小为 32.0GB；
- 物理机的操作系统为 Windows 10；
- 虚拟机为 VMware Workstation 16 Pro；
- 网络模拟器选用 GNS3 2.2.31；
- 交换机为 Ethernet Switch；
- 防火墙为 Cisco ASAv9.17.1；
- 路由器为 Cisco IOL Router15.7；
- PC1 和 PC2 为 Virtual PC Simulator, version 0.8.2。

17.2.2 实验设计

根据网络拓扑（图 17-1）完成网络设备的选型：2 台防火墙选用 Cisco ASAv9.17.1 系列防火墙，3 台路由器选用 Cisco IOL Router15.7 系列路由器。根据前面的组网需求，本实验设计如下。

1. IP 地址规划

网络有 5 个不同的子网，网络地址分别为 202.1.1.0/24、192.168.1.0/24、202.2.2.0/24、172.16.2.0/24 和 192.168.241.0/24。各设备接口的 IP 地址/掩码如表 17-1 所示。

表 17-1 各设备接口的 IP 地址/掩码

设备	接口	IP 地址/掩码
ASA1	G0/0	202.1.1.100/24
ASA1	G0/1	192.168.1.100/24
ASA2	G0/0	202.2.2.2/24
ASA2	G0/1	172.16.2.1/24
R1	E0/0	202.1.1.1/24
R1	E0/2	202.2.2.1/24
R2	E0/0	172.16.2.2/24
R2	E0/1	192.168.241.136/24（DHCP）
PC1	E0	192.168.1.188/24
Server（R3）	E0	192.168.1.1/24
PC2	E0	192.168.241.188/24
TEST PC（Cloud）	Eth1	VMnet8 IP：192.168.241.131/24

2. 路由设计

在路由器 R2 上配置一条到达 192.168.1.0/24 的静态路由（下一跳 IP 地址为 172.16.2.1）。

在防火墙 ASA1 上配置一条下一跳 IP 地址为 202.1.1.1 的默认路由。

在防火墙 ASA2 上配置一条下一跳 IP 地址为 202.2.2.1 的默认路由，定义一条到达 192.168.241.0/24 的静态路由（下一跳 IP 地址为 172.16.2.2）。

在服务器 Server 上增加配置一条下一跳 IP 地址为 192.168.1.100 的默认路由。

3. IPsec VPN 设计

首先定义安全域：将 ASA1 和 ASA2 的接口 G0/0 定义为外部接口，安全级别采用默认值 0；将接口 G0/1 定义为内部接口，安全级别采用默认值 100。

在防火墙 ASA1 和 ASA2 之间配置站点到站点的 IPsec VPN，具体包括在防火墙 ASA1 和 ASA2 上分别开启并配置 ISAKMP 策略，加密算法选择 3DES，Hash 函数使用 SHA，认证使用预共享密钥并在两台路由器上设置相同的共享密钥。在防火墙 ASA1 和 ASA2 上定义需要加密的数据流，要求对 192.168.1.0/24 和 192.168.241.0/24 这两个网段之间的流量进行加密。为此，在两个防火墙上定义两条互为镜像的 ACL 规则。

17.2.3 操作步骤

下面给出本次实验的具体操作，一共包括 10 个步骤。

1. 按照图 17-1 所示的网络拓扑，在 GNS3 中新建一个空白的工程

根据网络拓扑，在 GNS3 中完成网络设备选型及连线，其中，R1 和 R2 及 Server 选用 Cisco IOL Router15.7，ASA1 和 ASA2 均选用 CiscoASAv9.17.1-1，PC1 和 PC2 均选用 VCPS（然后右击更换图标）。可以先不连线 TEST PC（具体连接方法见后面）。

此外，交换机 Switch1 和 Switch2 的所有接口均属于 VLAN1，无须专门配置。

2. 根据表 17-1，配置设备接口的 IP 地址

1）配置防火墙 ASA1 接口的 IP 地址

```
CISICOASA(config)hostname cxyASA1        //可将 cxy 改为你的名字拼音首字母
cxyASA1(config)#interface g0/0
cxyASA1(config-if)#nameif outside
INFO: Security level for "outside" set to 0 by default.
cxyASA1(config-if)#ip address 202.1.1.100 255.255.255.0
cxyASA1(config-if)#no shutdown
cxyASA1(config-if)#exit
cxyASA1(config)#interface g0/1
cxyASA1(config-if)#nameif inside
INFO: Security level for "inside" set to 100 by default.
cxyASA1(config-if)#ip address 192.168.1.100 255.255.255.0
cxyASA1(config-if)#no shutdown
cxyASA1(config-if)#exit
```

2）配置防火墙 ASA2 接口的 IP 地址

CISICOASA(config)hostname cxyASA2
cxyASA2(config)#interface g0/0
cxyASA2(config-if)#nameif outside
INFO: Security level for "outside" set to 0 by default.
cxyASA2(config-if)#ip address 202.2.2.2 255.255.255.0
cxyASA2(config-if)#no shutdown
cxyASA2(config)#interface g0/1
cxyASA2(config-if)#nameif inside
INFO: Security level for "inside" set to 100 by default.
cxyASA2(config-if)#ip address 172.16.2.1 255.255.255.0
cxyASA2(config-if)#no shutdown

3）配置路由器 R1 接口的 IP 地址

R1(config)#hostname cxyR1
cxyR1(config)#interface e0/0
cxyR1(config-if)#ip address 202.1.1.1 255.255.255.0
cxyR1(config-if)#no shutdown
cxyR1(config-if)#exit
cxyR1(config)#interface e0/2
cxyR1(config-if)#ip address 202.2.2.1 255.255.255.0
cxyR1(config-if)#no shutdown
cxyR1(config-if)#exit

4）配置路由器 R2 接口的 IP 地址

R2(config)#hostname cxyR2
cxyR2(config)#interface e0/0
cxyR2 (config-if)#ip address 172.16.2.2 255.255.255.0
cxyR2 (config-if)#no shutdown
cxyR2 (config-if)#exit
cxyR2 (config)#interface e0/1
cxyR2 (config-if)#ip address 192.168.241.136 255.255.255.0
cxyR2 (config-if)#no shutdown
cxyR2 (config-if)#exit

5）配置路由器 R3 接口的 IP 地址

R3(config)#hostname cxyServer
cxyServer (config)#interface e0/0
cxyServer (config-if)#ip address 192.168.1.1 255.255.255.0

cxyServer (config-if)#no shutdown

把 R3 的图标改为服务器的图标：在 GNS 拓扑图中右击 R3，在弹出的快捷菜单中选择"change symbol"命令，然后在打开的对话框中单击"custom symbols"项即可找到合适的图标。

6）配置 PC 的 IP 地址及网关

对 PC1、PC2 设置 IP 地址及网关（TEST PC 先不用设置，具体操作见后文）。

PC1> ip 192.168.1.188/24 192.168.1.100

PC2> ip 192.168.241.188/24 192.168.241.136

7）测试直连网段的连通性

cxyASA1(config)#ping 202.1.1.1

Type escape sequence to abort.

Sending 5, 100-byte ICMP Echos to 202.1.1.1, timeout is 2 seconds:

!!!!!

Success rate is 100 percent (5/5), round-trip min/avg/max = 1/2/10 ms

cxyASA2(config)#ping 192.168.2.1

Type escape sequence to abort.

Sending 5, 100-byte ICMP Echos to 192.168.2.1, timeout is 2 seconds:

!!!!!

Success rate is 100 percent (5/5), round-trip min/avg/max = 1/2/10 ms

cxyR1#ping 202.1.2.100

Type escape sequence to abort.

Sending 5, 100-byte ICMP Echos to 202.1.2.100, timeout is 2 seconds:

!!!!!

Success rate is 100 percent (5/5), round-trip min/avg/max = 1/2/4 ms

cxyServer#ping 192.168.1.100

Type escape sequence to abort.

Sending 5, 100-byte ICMP Echos to 192.168.1.100, timeout is 2 seconds:

.!!!!

Success rate is 80 percent (4/5), round-trip min/avg/max = 1/1/4 ms

3. 配置静态路由

cxyASA1(config)#route outside 0 0 202.1.1.1

cxyASA2(config)#route outside 0 0 202.2.2.1

cxyASA2(config)#route inside 192.168.241.0 255.255.255.0 172.16.2.2 1

cxyServer (config)#ip route 0.0.0.0 0.0.0.0 192.168.1.100

cxyR2(config)#ip route 192.168.1.0 255.255.255.0 172.16.2.1

注意：R1 上不要设置任何静态路由，添加静态路由后，总公司、分公司及互联网三个区域内部的主机可以互通，但不同区域之间的主机是 ping 不通的。请用 ping 命令自行验证，例如：

PC2>ping 192.168.1.1

```
192.168.1.1 icmp_seq=1 timeout
192.168.1.1 icmp_seq=2 timeout
192.168.1.1 icmp_seq=3 timeout
192.168.1.1 icmp_seq=4 timeout
192.168.1.1 icmp_seq=5 timeout
```

4. 配置站点到站点（site to site）的 IPsec VPN

1）在 ASA1 上配置 ikev1 策略

```
cxyASA1 (config)#crypto ikev1 policy 1                          //创建序号为 1 的 ikev1 策略
cxyASA1 (config-ikev1-policy)#authentication pre-share          //认证采用预共享密钥
cxyASA1 (config-ikev1-policy)#encryption aes                    //AES 加密
cxyASA1 (config-ikev1-policy)#hash sha                          //Hash 算法选用 SHA
cxyASA1 (config-ikev1-policy)#group 14                          //Diffie-Hellman 组 ID 为 14
cxyASA1 (config-ikev1-policy)#lifetime 86400                    //SA 生存期 86400s
cxyASA1 (config-ikev1-policy)#exit
cxyASA1 (config)#crypto ikev1 enable outside                    //在外部接口启用 ikev1
```

2）在 ASA1 上配置 IPsec 隧道

```
cxyASA1(config)#tunnel-group 202.2.2.2 type ipsec-l2l           //202.2.2.2 为隧道对端 IP 地址
cxyASA1(config)#tunnel-group 202.2.2.2 ipsec-attributes
cxyASA1(config-tunnel-ipsec)#ikev1 pre-shared-key cxyuserpass
//指定预共享密钥为 cxyuserpass
cxyASA1(config-tunnel-ipsec)#exit
```

3）在 ASA2 上配置 ikev1 策略

```
cxyASA2 (config)#crypto ikev1 policy 1
cxyASA2 (config-ikev1-policy)#authentication pre-share
cxyASA2 (config-ikev1-policy)#encryption aes
cxyASA2 (config-ikev1-policy)#hash sha
cxyASA2 (config-ikev1-policy)#group 14
cxyASA2 (config-ikev1-policy)#lifetime 86400
cxyASA2 (config)#crypto ikev1 enable outside
```

4）在 ASA2 上配置 IPsec 隧道

```
cxyASA2 (config)#tunnel-group 202.1.1.100 type ipsec-l2l
//202.1.1.100 为隧道对端 IP 地址
cxyASA2 (config)#tunnel-group 202.1.1.100 ipsec-attributes
cxyASA2(config-tunnel-ipsec)#ikev1 pre-shared-key cxyuserpass
cxyASA2(config-tunnel-ipsec)#exit
```

5）定义名为"cxytransformSet"的 IPsec 转换集

cxyASA1(config)#crypto ipsec ikev1 transform-set cxytransformSet esp-aes esp-sha-hmac
cxyASA2(config)#crypto ipsec ikev1 transform-set cxytransformSet esp-aes esp-sha-hmac

6）定义感兴趣的流量（需要 VPN）

cxyASA1(config)#access-list l2l_list extended permit ip 192.168.1.0 255.255.255.0 192.168.241.0 255.255.255.0
cxyASA2(config)#access-list l2l_list extended permit ip 192.168.241.0 255.255.255.0 192.168.1.0 255.255.255.0

7）定义加密映射

cxyASA1 (config)#crypto map cxymap 1 match address l2l_list
cxyASA1 (config)#crypto map cxymap 1 set peer 202.2.2.2
cxyASA1 (config)#crypto map cxymap 1 set ikev1 transform-set **cxytransformSet**
cxyASA2 (config)#crypto map cxymap 1 match address l2l_list
cxyASA2 (config)#crypto map cxymap 1 set peer 202.1.1.100
cxyASA2 (config)#crypto map cxymap 1 set ikev1 transform-set cxytransformSet

8）在接口上应用加密映射

ASA1(config)#crypto map cxymap interface outside
ASA2(config)#crypto map cxymap interface outside

9）保存配置

cxyASA1 (config)#write memory
cxyASA2 (config)#write memory

10）调试 IPsec VPN

（1）从 PC2 上 ping192.168.1.1。

PC2>ping 192.168.1.1
84 bytes from 1 92.168.1.1 icmp_seq=1 ttl=254 time=4.621ms
84 bytes from 1 92.168.1.1 icmp_seq=2 ttl=254 time=2.693ms
84 bytes from 1 92.168.1.1 icmp_seq=3 ttl=254 time=3.237ms
84 bytes from 1 92.168.1.1 icmp_seq=4 ttl=254 time=3.586ms
84 bytes from 1 92.168.1.1 icmp_seq=5 ttl=254 time=3.711ms

（2）从 PC1 上 ping 192.168.241.188。

PC1>ping 192.168.241.188
192.168.241.188 icmp_seq= 1 timeout
84 bytes from 192.1 68.241.188 icmp_seq=2 ttl=63 time=5.341ms
84 bytes from 192.1 68.241.188 icmp_seq=3 ttl=63 time=2.660ms

84 bytes from 192.1 68.241.188 icmp_seq=4 ttl=63 time=4.385ms

84 bytes from 192.1 68.241.188 icmp_seq=5 ttl=63 time=1.952ms

（3）在 ASA1 上查看 ISAKMP。

ASA1(config)#show crypto isakmp sa
IKEv1 SAs:
Active SA: 1
　Rekey SA: 0 (A tunnel will report 1 Active and 1 Rekey SA during rekey)
Total IKE SA: 1
1　IKE Peer: 202.2.2.2
　　Type: L2L   Role: initiator
　　Rekey: no   State: MM_ACTIVE
There are no IKEv2 SAs

（4）在 ASA2 上查看 ISAKMP。

ASA2(config)#show crypto isakmp sa
IKEv1 SAs:
Active SA: 1
　Rekey SA: 0 (A tunnel will report 1 Active and 1 Rekey SA during rekey)
Total IKE SA: 1
1　IKE Peer: 202.1.1.100
　　Type: L2L   Role: responder
　　Rekey: no   State: MM_ACTIVE
There are no IKEv2 SAs

（5）在 ASA1 上查看 VPN 加密信息。

ASA1(config)#show crypto ipsec sa
interface: outside
　Crypto map tag: cxymap, seq num: 1, local addr: 202.1.1.100
　　access-list l2l_list extended permit ip 192.168.1.0 255.255.255.0 192.168.241.0 255.255.255.0
　　local ident (addr/mask/prot/port): (192.168.1.0/255.255.255.0/0/0)
　　remote ident (addr/mask/prot/port): (192.168.241.0/255.255.255.0/0/0)
　　current_peer: 202.2.2.2

　　#pkts encaps: 14, #pkts encrypt: 14, #pkts digest: 14
　　#pkts decaps: 14, #pkts decrypt: 14, #pkts verify: 14
　　#pkts compressed: 0, #pkts decompressed: 0
　　#pkts not compressed: 14, #pkts comp failed: 0, #pkts decomp failed: 0
　　#pre-frag successes: 0, #pre-frag failures: 0, #fragments created: 0
　　#PMTUs sent: 0, #PMTUs rcvd: 0, #decapsulated frgs needing reassembly: 0

#TFC rcvd: 0, #TFC sent: 0
#Valid ICMP Errors rcvd: 0, #Invalid ICMP Errors rcvd: 0
#send errors: 0, #recv errors: 0

local crypto endpt.: 202.1.1.100/0, remote crypto endpt.: 202.2.2.2/0
path mtu 1500, ipsec overhead 74(44), media mtu 1500
PMTU time remaining(sec): 0, DF policy: copy-df
ICMP error validation: disabled, TFC packets: disabled
current outbound spi: F7D7DB6E
current inbound spi: F7A2A84E
inbound esps as:
  spi: 0xF7A2A84E(4154632270)
    SA State: active
    transform: esp-aes esp-sha-hmac no compression
    in use settings={L2L, Tunnel, IKEv1, }
    slot: 0, conn_id: 6, crypto-map: cxymap
    sa timing: remaining key lifetime (kB/sec): (3914998/28617)
    IV size: 16 bytes
    replay detection support: Y
    Anti replay bitmap:
      0x00000000 0x00007FFF
outbound esps as:
  spi: 0xF7D7DB6E(4158118766)
    SA_State: active
    transform: esp-aesesp-sha-hmacnocompression
    in use settings={L2L, Tunnel, IKEv1, }
    slot: 0, conn_id: 6, crypto-map: cxymap
    sa timing: remaining key lifetime (kB/sec): (3914998/28617)
    Iv size: 16 bytes
    replay detection support: Y
    Anti_replay bitmap:
      0x00000000 0x00000001

(6) 在 R2 上使用不同的源 IP 地址 ping Server。

cxyR2#ping 192.168.1.1 source 192.168.241.136
Type escape sequence to abort.
Sending 5, 100-byte ICMP Echos to 192.168.1.1, timeout is 2 seconds:
Packet sent with a source address of 192.168.241.136
!!!!!

Success rate is 100 percent(5/5), round-trip min/avg/max=2/3/4ms
cxyR2#ping 192.168.1.1 source 172.16.2.2
Type escape sequence to abort.
Sending 5, 100-byte ICMP Echos to 192.168.1.1, timeout is 2 seconds:
Packet sent with a source address of 172.16.2.2
Success rate is 0 percent (0/5)

可见，在 R2 上使用不同的源 IP 地址 ping Server，得到不一样的结果。请解释其原因。

下面对 Cloud1 进行设置，以实现 PC2 访问互联网，以及通过分公司 TEST PC 上的浏览器访问总公司的 Web 服务器 Server。

5. 设置 Cloud1

从交换机 Switch2 连接到 Cloud1 的 Eth1 接口，该接口连接到虚拟交换机 VMnet8，即工作于 NAT 模式。

6. 设置 PC2 可以访问互联网

1）在 R2 上设置默认路由

在 R2 上设置到互联网的流量转向 VMnet8 网络的网关（如 192.168.241.2，该地址一定是虚拟机上 VMnet8 的网关 IP 地址）：

cxyR2(config)#ip route 0.0.0.0 0.0.0.0 192.168.241.2

2）为 PC2 设置 DNS 服务器

在 PC2 上设置 DNS 服务器地址：

PC2> ip dns 8.8.8.8

3）访问电子工业出版社域名

PC2> ping www.phei.com.cn

84 bytes from 14.215.177.38 icmp_seq=1 ttl=128 time=36.534ms
Redirect Network, gateway 192.168.241.136→192.168.241.2
84 bytes from 14.215.177.38 icmp_seq=1 ttl=128 time=46.517ms
84 bytes from 14.215.177.38 icmp_seq=2 ttl=128 time=53.744ms
84 bytes from 14.215.177.38 icmp_seq=3 ttl=128 time=55.953ms
84 bytes from 14.215.177.38 icmp_seq=4 ttl=128 time=39.025ms
84 bytes from 14.215.177.38 icmp_seq=5 ttl=128 time=81.377ms

至此，GNS3 中的 PC 可以访问互联网。

7. 在 R3 路由器上启用 HTTP

cxyServer(config)#ip http server   //启用 HTTP 服务
cxyServer(config)#ip http authentication local   //通过本地认证
cxyServer(config)#username cxylab privilege 15 password 0 cxylab   //用于登录 HTTP

8. 物理机设置

（1）查询 VMnet8 接口的编号。

先以管理员身份运行"cmd"命令，然后通过"route print"命令查询 VMnet8 接口的编号，如下面加黑体字所示。

```
C：\Windows\system 32>route print
-------------------------------------------------------------------------------
接口列表
14...98 8f e0 61 18 2b... ... Intel(R) Ethernet Connection (16) I219-V
23...02 00 4c 4f 4f 50... ... Microsoft KM-TEST 环回适配器
17...e2 0a f6 79 6e e1... ... Microsoft Wi-Fi Direct Virtual Adapter
22...f2 0a f6 79 6e e1... ... Microsoft Wi-Fi Direct Virtual Adapter # 2
05...00 50 56 c0 00 01... ... VMware Virtual Ethernet Adapter for VM net1
24...00 50 56 c0 00 08... ... VMware Virtual Ethernet Adapter forVMnet8
13...00 e0 6c 38 d4 e4... ... Realtek USB GbE Family Controller
66...00 50 56 c0 00 10... ... VMware Virtual Ethernet Adapter for VMnet16
08...e0 0a f6 79 6e e1... ... Realtek RTL8852BE Wi-Fi 6 802.11ax PCIe Adapter
 1          Software Loopback Interface 1
-------------------------------------------------------------------------------
```

（2）增加到 GNS3 的静态路由。

```
C:\Windows\system32>route add 192.168.1.0 MASK 255.255.255.0 192.168.241.136 METRIC 3 IF 24              //24 是前面查询到的 VMnet8 的接口编号
```

（3）在物理机上访问 Server。

先用 ping 命令测试：

```
C:\Windows\system32>ping 192.168.1.1 -S 192.168.241.1
```

再在浏览器地址栏内输入 http://192.168.1.1，按提示输入用户名和密码（例如，cxylab 和 cxylab），即可访问。

9. 设置虚拟机（TEST PC，如 Windows 10 x64）

（1）虚拟机的网络适配器设置为 NAT 模式（一般默认就是 NAT 模式）。

（2）以管理员身份运行"cmd"命令，增加到 GNS3 的静态路由。

```
C:\Windows\system32>route add 192.168.1.0 MASK 255.255.255.0 192.168.241.136 METRIC 3 IF 14              //14 为 VMnet8 的接口编号
```

（3）在虚拟机 TEST PC 的浏览器中输入 http://192.168.1.1，按提示输入用户名和密码。

10. 抓包分析

请自行抓包并进行分析，以加深对 IPsec VPN 的理解。例如，在虚拟机中运行"ping

192.168.1.1 –t"命令，然后抓包并进行分析，如图 17-2 所示。

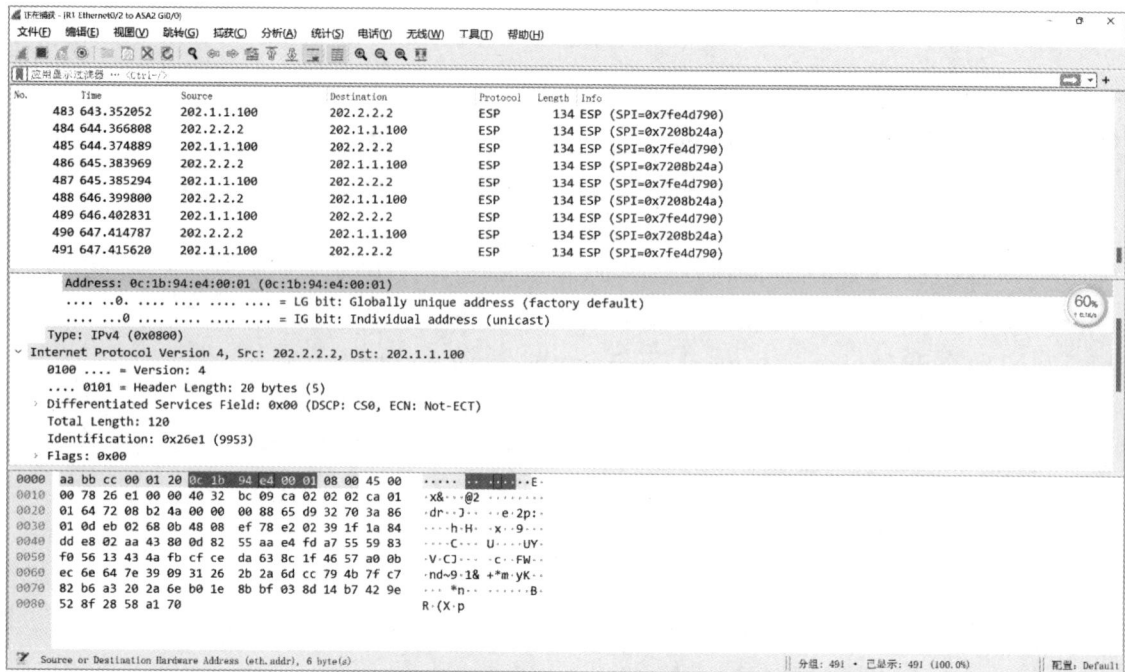

图 17-2　IPsec VPN 抓包分析